Simplest Universe Theory II

Donald Bovee

Simplest Universe Theory II

First revised edition, January 2018.
Second revised edition, November 2018.
Third revised edition, November 2020.
Fourth revised edition, April 2021.
Fifth revised edition, August 2021.
Sixth revised edition, March 2022.
Seventh revised edition, February 2023.
Eighth revised edition, September 2023.
Ninth revised edition, May 2024.

Please visit http://www.simplestuniverse.com for new material and updates.

Published by Fanton Press

ISBN: 978-0-9834915-4-5

After many years of wanting to live in France, I arrived in 2006. I began to think about physics for the first time in my life. Within two months, I had found that a single assumption about the beginning of the universe would lead to a complete theory of the universe. Since that extraordinary, but difficult, birth of this new theory, whenever I have lived in France, the theory has moved forward. Whenever I have lived elsewhere, it has stagnated.

I don't know why France has been the muse which has allowed me to unlock the mystery of why our universe exists and its laws. I only know that without France this theory would not exist.

I dedicate this theory to France.

Table of Contents

Chapter 3

Chapter 4

Chapter 5

Chapter 6

Chapter 21

Preface

Simplest Universe Theory II, now in its ninth revision, started out as a complete rewrite of *Simplest Universe Theory*, published in 2011. The first book attempted to explain a theory about the universe in a qualitative manner. Much of that book is incorrect and none of it rises to the level necessary for a description of a new theory of physics. Since 2011, I have learned the importance of describing the theory in both a qualitative and quantitative manner.

The standard model of physics has become comfortable with what are largely mathematical concepts, such as singularities and extra dimensions. They have their place as ways to make a physical theory easier to calculate, but they cause trouble when they are regarded as being real. There is no fourth dimension, a fact which has been verified by observation. A successful theory of physics should be able to explain why we observe only three dimensions.

There are no black holes in the sense of a collapse into a point in space. Large concentrations of matter do exist, often at the heart of galaxies, but their density is never greater than that of a pulsar, and, for concentrations of matter more massive than 2.3 solar masses, the density is less.

There are no quarks, nor are there W and Z bosons that mediate the weak force. This represents an abrupt change from the standard model. Yukawa's theory, in 1935, that the nucleus of an atom was held together by an unseen particle passing between the nucleons, became the centerpiece of the new physics. All forces were mediated by force-specific bosons. The boson model is not supported by the simplest universe theory. In calculating the branching ratios of the kaon, there are no signs of the involvement of either W or Z bosons.

Not surprisingly, classical physics remains unscathed in the new theory, though the interpretation of what is occurring sometimes changes. Gravitation is not the attraction of one object for another. Einstein had made that clear when he introduced general relativity in 1915. With the simplest universe theory it is now possible to explain what role space plays in gravitation, and why it appears that matter can attract matter, when in fact matter repels matter.

There is no quantum-classical limit, as was once espoused by Bohr. Classical theory is still an approximation of an underlying quantum reality, as Bohr imagined, but when we use classical theory to deal with the motion of an electron in an atom, it does not force electrons to jump through space without accelerating and decelerating. The results using a classical approach are the same as with quantum theory.

The distance between the particles from which space is made is sufficiently small that we can ignore quantum effects with gravity. We can treat gravitation as continuous in space. With the electrostatic force this is not the case. The traditional interpretation of Gauss's law does not work for the helium atom. At the same time, Coulomb's law works both for the very large and the very small, requiring adjustment only at distances equal to the diameter of a proton.

The simplest universe theory could be said to be axiomatic. A single assumption is used to provide the basis of the physics of our universe. Since all the forces and their interactions arise from this one assumption, the pieces of the theory are joined from the beginning. The difficulty of combining gravitation with electromagnetism disappears. The basis of the new theory is supplied in the appendix, so that the body of the book can concentrate on the physics of the new theory. The synopsis of the theory that is supplied in the introduction should nevertheless be read in order to understand the basis of the new theory.

That dark matter appeared late in the story of physics is unfortunate, because neither particle physics nor cosmology can be understood without including dark matter. The new theory requires dark matter from the beginning. It is one of only six stable particles that nature can build. From the point of view of the present standard model of particle physics, a particle of dark matter could be described as a bound state of an electron and a positron. The new theory makes explicit what dark matter is, how it can be found, and what its properties are.

In order to avoid repeatedly writing out "simplest universe theory," I refer to it as SUiTe, a semi-acronym suggested to me by Donna Calhoun. The theory is French, by which I mean that I had never thought of becoming a theoretical physicist until I first arrived in France in 2006. Since then, the theory has proceeded rapidly when I have been in France and slowly when I have lived elsewhere. I have no explanation for this.

The word *suite* in French indicates something that follows from something that came before. It can also be translated as "next." This theory follows from the extraordinary work done by the physics community. If I could compare the theory of physics to a jigsaw puzzle, all the pieces were there when I began, but some were in the wrong place. SUiTe supports the view held by Wigner (*The Unreasonable Effectiveness of Mathematics in the Natural Sciences*), and many others, regarding our ability to describe our universe with mathematics. In SUiTe, both the universe and the mathematics necessary to understand it were created simultaneously.

References

SUiTe is self-contained. For that reason, I have made little use of the available physics literature. When I have wanted to look something up, I have usually gone to Wikipedia (wikipedia.org) or Hyperphysics (hyperphysics.phy-astr.gsu.edu). Wikipedia was particularly useful for information about the elements and their isotopes, which I needed in order to develop a new nuclear shell model. For the masses and decays of particles, and for the most recent values of the constants of physics, I have used the Review of Particle Physics (pdg.lbl.gov).

I have also made use of four books. For wave theory, Maxwell's equations, and the properties of light, I have used Ditchburn, R.W. (1991) *Light*, Dover Publications, New York. For the original quantum theory, I have used Gautreau, R. and Savin, W. (1999) *Schaum's Outline of Theory and Problems of Modern Physics*, 2nd edition, McGraw-Hill. For example problems using the Schroedinger equation, I have used Holzner, S. (2009) *Quantum Physics for Dummies*, Wiley Publishing, Indianapolis. In order to better understand the standard model of particle physics, and for the Dirac equation, Fermi's Golden Rule, and the Feynman calculus, I have used Griffiths, D. (2008) *Introduction to Particle Physics*, 2nd edition, Wiley-VCH, Weinheim.

Acknowledgments

I would like to thank friends and family for being supportive of a project that no one in their right mind should ever attempt. In particular, I have benefitted from many conversations with Anthony Misch about the possible changes a complete theory of the physical universe would bring to weighty topics such as time and existence. I took my more mathematically oriented questions to John Sylvester, Peter Schmid, and Bernard Dekonick, who did what they could to help me. All gave good advice, and all were kind enough not to give the most obvious advice—that perhaps physics should be left to physicists. Donna Calhoun has commented on drafts of the theory.

Anthony Misch provided the cover and photographed the particles that SUiTe requires, created with plastic balls and wire in a way similar to models of molecules that are created in chemistry. Victoria Scott has helped me with both grammar and spelling on earlier editions. Those sections of the present text without errors are probably sections that remained unchanged from earlier versions that had been improved by her.

The many remaining errors in the book are my own.

Introduction

For the most part, the chapters follow the chronology of physics: Newton before Maxwell, Einstein before Bohr. In chapter 11, I introduce new work on the branching ratios of the kaon. In chapter 12, I show how this work is not unique to the kaon by extending the same methods to the tau. The purpose of moving this ahead of Bohr and the development of the quantum theory is to prepare the reader for a different way of doing physics by providing accurate theoretical estimates that cannot be derived from the standard model. From the Dirac equation to the present, the simplest universe theory creates a separate path that rarely uses the methods of the standard model of particle physics.

To some extent, each chapter is independent of the others. If you want to know the new theory of nuclear shells, there is no need to read earlier or later chapters.

I have put the more difficult and philosophical elements of the simplest universe theory in the appendix. There I start from the beginning of the multiverse, explain how it separates into universes, and continue until there is the usual space, matter, energy, and time that we find in our universe. The theory follows from a single initial assumption. This is the reason for calling the theory the *simplest* universe theory. There is nothing external to that initial assumption.

The present theory of physics resulted from the search to explain observations and experiments in the most concise manner possible, using mathematics. Often, phenomena that didn't initially appear to be related nevertheless followed the same mathematical pattern. Laws and principles were discovered that allowed physicists to predict the results of observations or experiments. This method has been successful in many of its parts but has led to contradictions within the overall theory. On the one hand, the method employed here avoids those contradictions; on the other hand, a model that depends on a single initial assumption is unlikely to be able to duplicate the complexity that we see around us. The following is a synopsis of the simplest universe theory.

Simplest Universe Theory

The multiverse is created from nothing in such a way that it must be able to return to nothing.

Ideally, any complete theory of the universe (or multiverse) should start from nothing. If it did not, it would be, at best, a complete theory subject to some set of initial conditions. It would then be necessary to explain where the initial conditions came from.

The second part of the above assumption, the part that requires the multiverse to be able to return to nothing, is surprisingly constraining. Any solution to this part of the initial assumption will require integer arithmetic. For every positive entity that is used to construct the universe, there must also be a negative entity with which it can combine so that both disappear. Such a multiverse is quantized from the beginning. But even that is not enough. Positive and negative entities must also attract each other. If positive entities attract positive entities, that will prevent positive and negative entities from coming together. Positive entities must either ignore other positive entities or repel them. Ignoring other positive entities leads to a universe that does not necessarily disappear. We have to choose repulsion. With that choice, we have taken the first step toward building a theoretical universe that is similar to our own.

How the single assumption leads from a need to create positive and negative entities to positive and negative forces, and from there to fundamental particles, and finally to combinations of the fundamental particles, is described in detail in the appendix. Two of these combinations, each one constructed from two fundamental particles, are responsible for space. Since they are constructed from two fundamental particles, I will call them *duals*. The duals are not in space; rather, they are space.

A particle of space has no clear boundaries because it is constructed of forces. The three fundamental forces of this model act along lines and do not emanate in all directions from a specific location, as is the case with a charged particle in the standard model of physics. Matter is also constructed of forces, and it is useful to have a definition for the location of a particle of matter. The location of a matter-particle is the locus of the bundle of forces which constitute the particle. Although this provides a location, it should not be forgotten that the forces extend beyond any sphere that can be drawn around this locus of forces. The extension of these forces gives matter wave-like properties.

The particles of space in each universe in the multiverse are constructed from the two weaker of the three fundamental forces in that region of the multiverse. The strength of a fundamental force is determined by its scarcity. Each fundamental force has a positive and a negative. There are 24 versions of space that can be constructed in the multiverse. There is a very large number of universes in the multiverse but not an infinite number of universes. If there were an infinite number of universes in the multiverse, it would be impossible for the multiverse to end, failing to meet the requirement of the one initial assumption.

Because the multiverse started from nothing—without time, energy, matter, forces, space, laws, principles, or mathematics—there is no part of physics that is external to the simplest universe theory (SUiTe).

1.1 Time

It is possible to speak of our universe beginning about 13.8 billion years ago. The length of a second, or its equivalent, what I call physics time, is not only unique to each universe, but the length of these seconds change as each universe expands. In the future, 10 billion years from now, our universe will appear to be older than 24 billion years, because seconds are getting shorter. Into the past, 10 billion years ago, our universe was less than 4 billion years old. This is time based on inertial mass, physics time. Inertial mass depends upon the density of space. The density of space changes with velocity, in a gravitational field, and continues to decrease as the universe ages.

There is another sort of time which we associate with biology. Something has to be born before it can die. The beginning of the multiverse was the birth of this time. The initial assumption requires the multiverse to end. This form of time will end.

In our own universe, even though counting seconds is not reliable because the length of a second depends upon the expansion of space, the biological sort of time is reliable and is common to all universes. The lightest elements will appear first, along with stars that use only the lightest elements for fuel. The heavier elements will appear later. Matter will be equally dispersed at the beginning and will become more and more concentrated as the universe grows older. Our own universe is slowly becoming a collection of small island universes disconnected from their super-cluster neighbors.

Physics time, like space, can be measured in two ways: as the number of particles of space, or in terms of the speed of light. If space is measured as the number of particles of space between two points, than a second can be a measure of the time it takes for light to travel a distance of a certain number of particles of space. If SUiTe is true, the early universe, at high redshift, will seem to have created galaxies too early, but using a measure of the second in terms of a speed of light determined by the number of particles of space, a redshift of 14 would correspond to 15 of our seconds today. These ideas will be explored in chapter 4.

I.2 Particles of Space

In our universe, the electrostatic force is the strongest. The second strongest force is the weak force, which has some properties in common with the weak force of the standard model of physics. I have named the force that is even weaker than the weak force the dense force because the particles of space corresponding to the dense force are far more common than those corresponding to the weak force.

The definition of a fundamental particle in SUiTe is different than that of a fundamental particle in physics. In SUiTe, a fundamental particle is made of three forcelines from the three fundamental forces joined at right angles to each other. Because each force can be either positive or negative, eight such particles can be created. The fundamental particles do not take up space, but when two of these fundamental particles come together to form a dual particle, their inability to completely overlay each other creates space.

The dense-force particles of space (d-duals) in our universe are made from two fundamental particles that attract each other with the electrostatic and weak forces, but both of these fundamental particles are dense-force negative. The two dense-force negative lines of the dense-force dual must align themselves parallel to each other. The two other forcelines, which are the positive and negative of their forces, combine. The distance between these two parallel dense-force forcelines represents the smallest distance possible in our universe. This distance is much less than 10^{-27} meters, the approximate distance between d-duals.

The second particle of space is the weak-force particle, in which two negative forcelines of the weak force are parallel. I call this the weak-dual, or w-dual. Like the d-dual, the two parallel weak force forcelines hold apart the two fundamental particles from which the w-dual is made. The weak force is far stronger than the dense force, meaning that it is far less common, and the distance between the two fundamental particles of the w-dual is much greater than in the d-dual. The distance within the d-dual is the ultimate source of space. The number of d-duals per cubic meter is 10^{79} or more. The distance between the two fundamental particles that create a w-dual allows the w-dual to interact with the d-duals of space. One can think of the d-dual as creating the space in which the w-duals are able to pivot.

The two fundamental particles that create a w-dual have their attractive dense-force forcelines spread apart even though they are attracted to each other. The far more numerous d-duals interact with the spread-apart dense-force forcelines of the w-dual, causing the w-dual to spin. W-duals do not spin with any preferred direction. When w-duals spin in the same direction, they are the basis of magnetism. The linear ratio of d-duals to w-duals is proportional to the speed of light, because for a photon, which is made up of w-duals, to pass through space, it is necessary to rotate the w-duals such that the weak-force forcelines are perpendicular to the direction of motion of the photon. The d-duals resist this motion, slowing the rotation of the w-duals. The time that the rotation takes is proportional to the number of d-duals involved, which is proportional to the linear ratio of d-duals to w-duals in the universe and is proportional to c, the speed of light.

Space is neither continuous nor empty. It is created of particles. The d-duals, because of their repulsion for each other, form a three-dimensional lattice. The w-duals also form a similar lattice. A photon is a moving compression of w-duals. The need to rotate the w-duals in order for the photon to pass, since the w-duals are surrounded by d-duals, takes time. A photon passing through this double lattice distorts it. This distortion creates a transverse wave, but there is also the longitudinal wave created by the w-duals which is the source of light pressure.

Though it is possible to consider the w-duals as being particles in space, the d-duals should be considered as contiguous with each other. In that sense, the d-duals are space. Light is a phenomenon of the w-duals. For this reason, it makes little sense to talk of a d-dual as being in a surrounding space. The inability of two of the forcelines in the pair of particles that create a d-dual to completely overlay each other provides the minimum distance necessary for any mathematical description of space. Since all massive particles must be created with at least some particles of space included in their structure, it ensures that there will be no point particles in this theory.

The point in space that the locus of forces that a w-dual occupies cannot also be occupied by a d-dual. Similarly, the locus of the forces that constitute a non-compound particle of matter, such as an electron, cannot simultaneously be occupied by anything else, not even the d-duals. Every particle that is not a dual particle of space is a particle of matter. Protons and neutrons are each made of smaller particles of matter. They are compound particles. There are d-duals and w-duals inside them. The gravitational compression of neutrons in neutron stars will be resisted by the need to compress the duals of space inside the neutrons. The duals of space cannot be compressed infinitely, preventing the creation of black hole singularities.

D-duals repel d-duals, and w-duals repel w-duals. Space expands because of this repulsion. Matter is also created from fundamental particles. Matter attracts space. By attracting space, particles of matter retard the expansion of space. The inertial mass that we associate with a particle of matter is determined by the density of space surrounding the particle. As space expands, and as gravitation causes the creation of clusters and superclusters of galaxies, the number of particles of space interacting with particles of matter decreases, and the inertial mass of each matter-particle subsequently decreases.

The rest mass is dependent only on the number of space-attractive forcelines of matter particles and does not change with expansion. In this way, it is similar to the charge of a particle. It is the inertial mass which changes. It is possible to talk of a density of space because matter does not expand as space expands.

I.3 Particles of Matter

The initial region of the multiverse from which our universe sprang was able to create four complex particles. I have named them NEG, POS, neg, and pos. Each of the four is negative or positive with respect to the electrostatic force. The all upper or all lower case (NEG versus neg) indicates whether they attract or repel the particles of our space. Particles designated here in upper-case letters (NEG and POS) attract the duals of our space and have rest mass as a result. Every particle that is not one of the two duals that make up space has electric charge. Those particles that are neutral, such as the neutron, are constructed of an equal number of oppositely charged particles.

POS particles are identical in every way to NEG particles except for their charge. The same is true of pos and neg particles. An electron is constructed of three particles in the form neg/POS/neg. A positron is constructed of pos/NEG/pos. The three electrostatic forcelines associated with the three particles that make up the electron or the positron are parallel, and they are sufficiently close to one another that the electron appears to have only one negative forceline, and the positron has only one positive forceline. Both the electron and the positron have three weak-force forcelines held at right angles to the electrostatic forcelines. The placement of the weak-force forcelines explains the behavior of the electron when it is placed in an inhomogeneous magnetic field such as in a Stern-Gerlach apparatus. As explained in the appendix, the neg/POS/neg particles that formed the electron have become fused, such that the electron cannot be broken into its original constituent pieces. Both the electron and the positron are therefore non-compound matter particles.

Protons are constructed of four electrons and five POS particles in the shape of a three-ribbed sphere. Two of the ribs have electrons in the middle, while the third rib has two electrons separated by a POS particle. The antiproton is constructed by replacing electrons with positrons and POS particles with NEG particles.

The electron neutrino is constructed of a NEG and a POS particle. As can be imagined, the replacement of NEG with POS and POS with NEG results in the same particle. The neutrino is therefore its own antiparticle. The muon neutrino and the tau neutrino are constructed of multiple electron neutrinos. Neither of them is their own antiparticle. Given sufficient energy, the electron neutrino can be broken, leaving free NEG and POS particles. These two particles can recombine with each other to reform an electron neutrino or with positrons and electrons to form antiprotons and protons. Because each broken neutrino provides one NEG particle and one POS particle, protons and antiprotons will be created in equal numbers.

The proton and antiproton, electron and positron, and the electron neutrino are all stable. The last of the six stable particles of our universe is dark matter. Like electron neutrinos, particles of dark matter are their own antimatter particles.

I.4 From Matter to Mass

In SUiTe, matter is conserved. The gravitational mass, the rest mass of matter, depends only on the number of attractive forcelines that matter has for space. It is similar to electric charge and does not change over time, even with the expansion of space. Inertial mass reflects the difficulty of moving matter against space. When space is compressed in front of a moving object or when space is contracted around a massive object because of gravitation, the density of space increases. Inertial mass is then dependent on the density of space. It is never less than the rest mass and is approximately equal to the rest mass at slow speeds and in weak gravitational fields. It is higher than the rest mass at relativistic speeds, or in strong gravitational fields.

The expansion of space, going into the past is in some ways similar to entering a gravitational well. If we compare a photon sent from the past to the same photon today, the one from the past has less energy in the same way that it would if the photon was sent from within a gravitational well. However, if we examine the characteristics of a star in the past, it is no different than today. Both inertial mass and seconds have changed in comparison with today, but this is due to the way that we measure them. Although the energy of a photon from the past has less energy than a similar photon today, in the past its energy with respect to all other energies was no different than similar relations now. The expansion of space results in a scaling change where, because fewer particles of space directly interact with matter as space expands, the inertial mass is increased in the past relative to today, resulting in lower energy photons from the past when measured in today's world of shorter seconds.

Chapter 1

Newton's Laws

1.1 A Body in Motion Stays in Motion

The first law requires that a body at rest remain at rest, and a body in motion remain in motion, unless either body is acted on by unbalanced forces. The rest-state of Newton is very difficult to define, given that the Earth is spinning and also orbiting the Sun. The solar system is rotating about the center of the Milky Way galaxy, which is moving in the direction of the Virgo supercluster, and so on. Nothing around us is at rest. It won't be possible to define an absolute rest-state until the chapter on special relativity.

The first law suggests the conservation of momentum. How does a body remain in motion in SUiTe, given that inertial mass can be considered as a form of friction between matter and space? Friction decelerates an object. A force, defined in Newton's second law, accelerates (or decelerates) an object. Friction is therefore a force. If an object moves through space without slowing down, there must be a countervailing force that offsets the friction of space. What is the source of this acceleration?

If an object is accelerated by some force, it compresses space in front of it. The compression of space at the front of the object causes the attractive forcelines of the object (space attracts matter) to tilt in the direction of motion in order to take advantage of the attraction of space due to the compression.

The smallest particles from which we are made, including those from which protons and neutrons are made—electrons, neutrinos, and POS particles—cannot be compressed any further by space. If all of them are under equal pressure from all sides by space, how would we know?

If the pressure due to space is not distributed evenly, we move. Newton's first law depends on this change in pressure. When the forcelines of a particle do not point outward in all directions equally, the pressure of space is no longer equal on all sides of the particle. Pressure is a force. The friction of space that results in the inertial mass of subatomic particles is also a force, and the two forces equalize. But they can only equalize when the particle is either at rest or in constant motion.

Constant motion causes the forcelines of the particle to point in the direction of motion, if only by an incredibly small amount, due to their attraction for the space which is compressed in front of the particle. Because more of the forcelines are tilting forward, the pressure of space from behind no longer attracts as many of the particle's forcelines as it did when the particle was at rest. The forcelines of the particle and of space partially cancel each other when they attract each other. With less of this canceling from behind due to the shift of the particle's forcelines toward the front, but no reduction in pressure from space from behind, the particle experiences a force from behind equal to the friction in front. If we could see the tilt in the forcelines, we would know the velocity of the particle relative to some maximum velocity where the forcelines of the particle are all pointing forward. This way of determining velocity is further developed in the chapter on special relativity.

1.2 Newton's Second Law

The second law is deceptively simple: $F = ma$, where m is mass and a is acceleration. This was apparent from our description of the first law. If we have a force, and we want to avoid acceleration, then we must have a second force in the opposite direction that prevents acceleration. Try to imagine the first law without the second. Imagine that I push an object, and then I stop pushing. It will keep moving at the speed it reached, while I was pushing (the first law). If I continue to push, the object has to then move faster and faster. For motion to continue unchanged when I am no longer applying a force, the object must accelerate whenever I do apply a force. The first law implies the second.

1.3 Newton's Third Law

Here, Newton talks directly about momentum and the conservation of momentum. If I am standing on frictionless ice and I push an object forward, then I will be pushed backward. The amount that I will be pushed backward depends on the difference between my mass and the mass of the object I pushed forward. If the object was not moving before I pushed it, then the momentum of the object moving forward is equal but of the opposite sign to my momentum moving backward. Let's think about this specifically with two neutral particles, neutrons. In this imaginary case, they will behave like ideal spheres, running straight into each other with the same speed. They will then have equal but opposite momentum. The sum of the momenta is zero.

In SUiTe, a neutron in motion has its forcelines angled in the direction of motion. The particles of space are compressed in front of the neutron, attracting the forcelines of the neutron. Two neutrons, both on a path where they should collide, don't actually touch each other. As they get close, the particles of space that are compressed in front of both of the neutrons start to repel each other.

When the compressed particles of space, which look a bit like pairs of chopsticks, begin to run into each other, they can only reduce their repulsion by turning parallel to each other. They are then perpendicular to the direction of motion of the two neutrons. Perpendicular forces do not affect each other in SUiTe. As the forcelines of the duals of space become perpendicular in front of the neutrons that are colliding, they no longer attract the forcelines of the two neutrons. Without the attraction, the forcelines of the two neutrons point in all directions equally. They are temporarily at rest. The neutrons no longer feel a pressure from space from behind that causes them to move.

The particles of space in between the two neutrons are compressed and perpendicular to the earlier direction of motion of the neutrons. Without further pressure from the neutrons, now that the neutrons are at rest, the compressed space expands. The particles of space have no mass, so all the energy that had been stored in the compressed space between the two neutrons is released, sending the neutrons backward with the same speed but in the opposite direction.

Space stores energy when it is compressed in front of a moving particle or in the form of a photon, a moving compression of space that is not associated with a particle. Because the structure of space is similar to a very tightly bound crystal where the distance between the nodes of the crystal is extremely small, the moving compression does not dissipate. The particles of space have no mass and are unable to store energy without motion. When motion ends, all of the energy temporarily trapped in the moving compression of space is given up driving the two neutrons apart. The conservation of momentum in a collision is equivalent to the conservation of kinetic energy when the kinetic energy cannot be stored by modifying the particles involved in the collision.

1.4 Newton's Law of Gravitation

Newton discovered a law that described the behavior of the planets. He hypothesized that all matter affected all other matter according to his law. Where *G* is the gravitational constant:

$$F_G = G\,{}^{m_1 m_2}/_{r^2}$$

Here m_1 and m_2 are the masses of the two objects that are interacting. r is the distance between them. This is an example of an inverse-square-law force because the force involved increases or decreases in accordance with the inverse-square of the distance between the two objects involved. The choice of r, *radius*, for distance has an additional meaning; the best way to describe the gravitational force that an object feels is with polar coordinates. For example, at the distance from the Sun to the Earth, any object would feel the same gravitational force as the Earth does. If there were two Earths or five Earths, each would still feel the same gravitational force as long as they were small in comparison to the Sun and at the same distance from the Sun.

In the previous section, when an object was in motion, space was compressed in front of the object, and the forcelines of the object were tilted in the direction of motion. Let's assume that an object is not moving. The forcelines that attract space are pointing out in all directions equally. The granularity of space is very fine, because space is constructed of d-duals, which may be about 10^{-26} meters apart, and w-duals, which we know to be about 10^{-18} meters apart. Both duals are attracted to 5 of the 9 particles from which a proton is constructed in SUiTe. Neutrons have the same 5 attracting particles, but one additional electron and a neutrino for 11 particles in total.

Neutrons and protons attract the particles of space. The density of space decreases with distance from matter. The density is dependent on the attractive forcelines of matter. Those forcelines do not weaken with distance, and point outward from matter in all directions equally if we assume that the matter object is either not moving, or moving slowly with respect to light. All the forcelines of a massive object will pierce a hollow sphere surrounding it. A similar sphere twice the distance from the object will still have the same number of attractive forcelines piercing the sphere, but only one-fourth as many per square meter. This creates the inverse-square law for the gravitational force.

Gravity is a central force, but different from electrostatic attraction. What drives gravitation is the change in the density of space. With two bodies in space, the contraction of space on a line between the two bodies is slightly greater than in any other direction. The forcelines of these two bodies tilt slightly toward the greater contraction of space that exists between them. This tilting of the forcelines changes the overall pressure of space, acting on an object in the same way as described for Newton's first law of motion. The uneven pressure of space on the two bodies causes them to move toward each other—the direction of the greatest contraction of space. There is no attraction of matter for matter, only the attraction of matter for space and of space for matter. The way that Einstein modeled the bending of space in four dimensions is mathematically equivalent to the contraction of space in three dimensions.

Chapter 2

Special Relativity

Einstein, Lorentz, Poincaré, and others tried, in different ways, to accommodate the Michelson-Morley experiment within the classical framework of physics, as it existed at the beginning of the twentieth century. The experiment proved that the speed of light did not depend on the direction in which the Earth was moving. This was a complete repudiation of the existing theory that light traveled in a luminiferous aether. This aether was considered necessary for a wave of light to propagate. All waves must have an underlying medium. The foremost aether model was that due to Lorentz. There were problems with his model. For one thing, he accepted that there was a luminiferous aether capable of shrinking matter. The shrinking of matter perfectly offset the expected aether affect, such that light would *appear* to travel at the same speed in all directions.

Albert Einstein threw out the idea of a luminiferous aether and replaced it with a principle, the principle of relativity. There is no absolute velocity but only relative velocity. Every particle in the universe may consider itself as not moving, and consider every other particle as moving around it. The invariance of Maxwell's laws describing electromagnetism for objects in constant motion, combined with an unchanging speed of light, resulted in changes to the physics of motion for objects moving at near light speeds. Neither time nor velocity would necessarily appear to be the same for observers at different locations describing the same event. These changes were described by Einstein in 1905 in what is now known as Einstein's theory of special relativity.

2.1 Fitzgerald-Lorentz Contraction

Fitzgerald was the first, in 1889, to attempt to save the idea of a luminiferous aether two years after the Michelson-Morley experiment by suggesting that the Michelson-Morley experiment had failed to detect the aether because the arms of the experiment had shrunk when pointed against the direction of the aether flow. The Michelson-Morley experiment had four arms at right angles to each other, and at the end of each arm was a mirror. The purpose was to measure the speed of light by splitting a beam of light, bouncing it off the mirrors, then joining the split beam to check for interference. Michelson and Morley found no interference. If there had been an aether effect, one of the two split beams would have taken longer to reach the detector than the other, creating an interference pattern that could be detected.

In 1818, Fizeau measured the velocity of light, both with and against a flow of water, and found that light was faster in the same direction as the flow of the water. Because of this experiment, it seemed reasonable that there would also be a slight correction in the measurement of light depending on whether or not it was flowing with the aether. I am belaboring this point because there is a luminiferous aether in SUiTe, though not of the type thought of at that time. Lorentz, in 1892, came to the same conclusion as Fitzgerald. The apparatus of Michelson and Morley had shrunk when it was pointed against the aether wind.

In SUiTe, the contraction of matter is apparent but not real. Our meter ruler does not reflect changes in the density of space. The ruler that nature uses counts the number of particles of space between any two points in space. The compression of space does not change the number of particles of space between two points, even though, because we can't see the compression of space, the apparent distance will appear to have changed. If we could see a spaceship moving close to the speed of light, it would appear to be shorter in the direction of motion, but for the people on the spaceship, it would not have shrunk at all. A ruler that is based on the number of particles of space between two points is necessary for some measurements in physics.

2.2 Transverse Doppler Shift

Ives and Stilwell measured the energy of light emitted by an excited atom perpendicular to the motion of the atom. If the energy was less than normally expected, we could describe this change in light energy as Doppler-shifted toward the infrared, a redshift. The frequency of light is redshifted, if it has less energy than expected, and blueshifted, a shift toward the ultraviolet, if it has more energy than expected.

I will describe the ideal experiment, though this was not the experiment that Ives and Stilwell performed, but was instead what they succeeded in measuring in a different and indirect manner.

An excited atom moves in a stream of excited atoms through a laboratory. To the right of the stream of atoms is a mirror, and to the left of the stream is a detector that will measure any light emitted by an excited atom that deflects off the mirror perpendicular to the flow of the stream. Given the reflection, there is no question that the light was emitted at a 90-degree angle relative to the motion of the stream of excited atoms. The energy of the light is reduced by a factor of exactly

$$\sqrt{1 - {}^{v^2}/_{c^2}}$$

where v is the speed of the atom.

If ${}^{v}/_{c} = \cos\theta$, then the above square root is equal to $\sin\theta$.

What is θ the angle of? We know, from the previous chapter, that the motion of every particle causes the forcelines of the particle to tilt in the direction of motion. When a particle is not in motion, the forcelines stick out in all directions, like pins sticking out of a spherical pincushion. When the particle is moving, the forcelines tilt toward the compression of space in front of the particle, because space is attractive to matter. When the particle is at rest, a forceline sticking out perpendicular to the future motion of the particle will have an angle with the future direction of motion of 90 degrees and a cosine equal to 0. As the particle moves, this forceline will tilt forward, going from 90 degrees to 0 degrees. At 90 degrees, the velocity is zero and ${}^{v}/_{c} = 0$. At 0 degrees, the velocity is equal to the speed of light. The angle of the forceline of the particle in motion that was initially perpendicular to motion when the particle was at rest is equal to $\cos^{-1}(v/c)$.

We can pull some more information out of the above square root. The units of a velocity squared are those of a pressure divided by a density. What if that pressure divided by density refers to the particles of space? What if c^2 is the maximum pressure divided by density that is possible for space? Then dividing by c^2 will normalize the pressure divided by the density of space associated with v^2. Thinking of a velocity squared as indirectly measuring either the compression or the contraction of space will be useful in the next chapter, on general relativity.

With the exception of electrons, positrons, and neutrinos, the vast majority of the mass of an object depends on the attraction of space for the object. This attraction creates both the rest mass and the gravitational mass of the object. In special relativity, the motion of an object creates a compression of the particles of space in the direction of motion. This compression of space should modify the inertial mass of an object no differently than the contraction of space surrounding an object due to gravitation. If this is true, the inertial mass of an excited atom, in a stream of excited atoms whose emitted light we are attempting to measure, depends on the compression of space that results from the motion of that atom. Now we can go to the formula for energy, derived from Newton's second law of motion that defines force. Joule is responsible for the definition of energy as force multiplied by the distance, *m*, through which the force is acting:

$$E = F \cdot m = kg \cdot m^2/s^2.$$

Time, in physics, is the repetitive motion of matter. If the inertial mass of an object increases due to changes in the pressure divided by the density of space, then seconds must be longer by the same proportion that the inertial mass increases. Distance does not change if we use nature's ruler, counting the number of particles of space between two points. The compression of space by matter will change mass (kg) and seconds in the same direction by the same proportion. Seconds are squared in the denominator in the equation for energy. Energy will therefore decrease. We know by exactly how much because we know how velocity affects the pressure divided by the density of space, and therefore the change in the inertial mass of the object that is in motion.

The transverse Doppler shift shows that the change in energy is dependent on the speed of the particle emitting the energy. It also explains how matter and space interact to create additional mass in the form of inertial mass which exceeds the rest mass when matter is in motion. The transverse Doppler shift is not dependent on a principle of relativity that suggests that the atom sees the emitted light being sent backward to the detector that exists in the laboratory frame. The light is being emitted by the atom at right angles to its motion, exactly as it appears to do in the laboratory frame. The compression of space by the atom that has increased the inertial mass of the atom has resulted in the emission of a photon of light with less energy than it would otherwise have had.

2.3 Mass Increases, Time Slows

The contraction of a relativistic spaceship in special relativity is apparent but not real, while the change to inertial mass is real. The compression of space in the direction of motion requires matter to interact with more particles of space. The electrostatic force is not carried by space and is not affected by the density of space, but from Newton's second law, we know that when we consider how the electrostatic force acts to accelerate a particle, we must include the mass of the particle. Because mass is in the numerator in the equation for force, force increases with an increase in mass. Acceleration, however, decreases as the length of the second increases, and seconds are squared in the denominator. This is why light emitted by a moving particle perpendicular to its motion has reduced energy.

The proportional change to both force and energy due to motion is $(1 - v^2/c^2)^{0.5}$. The proportional change to mass and to the length of a second is $(1 - v^2/c^2)^{-0.5}$. The apparent contraction of a ruler in motion is equal to the proportional change to energy.

With the exception of the contraction of matter in the direction of motion, all the changes due to motion are real and not apparent. In SUiTe, it is not possible for a particle moving near the speed of light to be considered as not moving, while the laboratory in which the particle is measured is moving in the opposite direction. Velocity is not relative, it is absolute.

2.4 A Particle at Rest in the Aether

In SUiTe, a particle that has mass can be at rest with respect to the particles of space. Space is not truly an aether, and flows of space can only occur if we come into contact with another universe that succeeds in breaking down the wall that we presently have with other universes. For most problems, we can treat space as unmoving. If we equip every particle of matter with a clock, something that Einstein does with general relativity, the particles with the fastest clocks are at rest with respect to space. Even though we may perceive those particles as moving, because they are moving with respect to us, they are in fact not moving.

2.5 Dilation of Time

A major problem with the principle of relativity is the need to allow each particle to be at rest while other particles around it, depending on their relative velocities, move. Perhaps if Einstein had lived long enough to appreciate how colliders, by accelerating pions to near the speed of light, had managed to increase the pion lifetime by many magnitudes, he would have understood that changes to mass and to time are real. There is no sense in stating that a pion that decays in a fraction of a microsecond is no different than a pion traveling at almost light speed that fails to decay in a second. The fast-moving pion survived substantially longer because it was moving rapidly, and for no other reason. The convoluted explanation for the Twin Paradox is equally pointless.

If the primary force that we are interested in is the electrostatic force, and that the electrostatic force, in a massless sense, does not change with the density of space, then we can understand that a change in the mass of a particle will have to change the acceleration of that particle by the electrostatic force. If an electron takes longer to move in an atom due to its increase in mass, then every process in our body will also take longer. The change in mass of the particles of our body due to the motion of our body will cause us to live a longer life relative to someone who is not in motion. Gravity also causes an increase in the density of space, and gravity also causes an increase in the lifetime of objects in the gravitational field. It is odd that these two similar results have never been associated with a potential change in the density of space.

2.6 Relative Velocity

Although there is such a thing as a particle at rest with respect to space, and although we could potentially measure every other particle's motion by referring to a particle at rest in space, it is much easier to work with relative velocity. We should not make the mistake of thinking that we are not moving and everything else is, but when we are measuring particles moving close to the speed of light, it is safe to assume that we are not moving, and that they are moving, because the subsequent errors are small.

The statement that something moving relative to me will have a slower clock compared to mine is not necessarily true. If both of us have atomic clocks and you are moving faster in an absolute sense than me, your clock will be slower no matter what the relative velocity says. In one way, special relativity is not different from general relativity. The changes in time and in mass depend solely on the change to the density of space, something common to both theories. But there is one important difference between special and general relativity; in special relativity, the change in the density of space has direction, and therefore the change in inertial mass also has direction. This is not the case with general relativity. This is the problem with the definition of mass itself. We can associate the rest mass with the gravitational mass, but whenever we talk about motion, we must add to that its change to its inertial mass. This includes the motion of a particle attempting to move in a gravitational well. Most of physics is only concerned with the inertial mass of objects. The gravitational field depends solely on the rest mass.

Chapter 3

General Relativity

Einstein's general relativity explains Newton's gravitation in terms of the deformation of space. Newton did not accept mutual attraction as the explanation of gravity, nor did he accept any other explanation. If we fail to choose nature's ruler that depends on the number of particles of space, and instead, use a ruler that is not dependent on the density of space, then we will have no choice but to expand space to four dimensions, as Einstein did. The movement of a particle in a gravitational field is then a unique projection onto three dimensions from a carefully defined four-dimensional space in which time has been transformed into a fourth spatial dimension.

This method saves the ruler that we wish to use at the expense of failing to see how nature acts. There are only three dimensions. To understand the physics of space, we need to calculate distance in terms of the number of particles of space between two points. The contraction or expansion of space, when it has no effect on the number of particles of space between two objects, does not change nature's ruler, but it does change our preferred ruler. We can't do physics using our preferred ruler without making mistakes.

3.1 The Einstein Factor

In physics today, there is nothing named the "Einstein factor." There is, however, the Lorentz factor equal to:

$$1 \Big/ \sqrt{1 - v^2/c^2}$$

The Lorentz factor gives us the increase in mass and the increase in the length of a second. Its inverse gives us the apparent contraction of matter. The Lorentz factor can be used to construct a matrix which allows us to compare our clock and our ruler with the clock and ruler that we perceive in a laboratory that is moving with respect to us. As stated in the previous chapter, a velocity squared has the units of a pressure divided by a density. If we take c^2 to be the maximum possible pressure divided by density that space can achieve, then v^2/c^2 is the normalized, unit-less pressure divided by density associated with v^2. To create an Einstein factor, we will need to find a measure that is derived from the gravitational constant and that has the units of a velocity squared.

The gravitational potential energy of a small particle of mass *m* at a distance *r* from a much larger source of mass *M* is:

$$G\,{}^{Mm}/_{r}$$

The acceleration of the same small particle *m* in the direction of the larger source due to gravitation at a distance of *r* is:

$$G\,{}^{M}/_{r^2}$$

The relationship between acceleration and a velocity squared is given by Newton as:

$$v^2 = 2ad$$

where *a* is acceleration and *d* is distance. This equation will provide us with the units of a velocity squared for the acceleration of a small particle in the gravitational field of a mass *M*.

$$2G\left({}^{M}/_{r^2}\right)r = 2G\left({}^{M}/_{r}\right)$$

Turning this into an equivalent of the Lorentz factor, but for gravity, yields:

$$\frac{1}{\sqrt{1 - {}^{2GM}/_{rc^2}}}$$

It is this value that I want to call the "Einstein factor." From the Einstein factor, it is possible to arrive at the Schwarzschild solution that describes how a single massive body contracts the space around it.

3.2 The Schwarzschild Solution

The distance r in the Schwarzschild solution is the distance to a test particle within the field of a source of gravitation from the center of that source. The test particle is considered to be too small to have any gravitational influence of its own. We want to compare a clock and a ruler outside of the gravitational field with a clock and a ruler at the location of the test particle. A gravitational field is a contraction of space that depends on the distance from the source of the field and the mass of that source. Matter attracts space, which causes space to be denser the closer it is to matter.

The Schwarzschild solution is in the form of a second derivative where time is treated as a distance in an idealized fourth dimension by multiplying time by c. The projection of the four dimensions of this spacetime on to the three dimensions in which we exist allows us to sketch the path that either light or matter will take through a gravitational field. In special relativity, changes in speed resulted in changes to the clock of the particle, but did not alter the path of a particle. Gravitation, by modifying space, appears to exert a force on a particle which alters its path as well as its clock. In special relativity, if the speed does not change, the clock does not change. It is only when we compare the clocks of objects moving at different speeds that we see the relative change in time. In general relativity, if the distance from the source of the gravitational field to the particle does not change, then its clock does not change. All particles at a given radius from the center of the object creating the gravitational field will have the same change in time.

The preferred way to write the Schwarzschild solution is:

$$c^2d\tau^2 = \left(1 - {}^{R_S}\!/_{r}\right)c^2dt^2 - \left(1 - {}^{R_S}\!/_{r}\right)^{-1}dr^2 - r^2d\theta^2 - r^2sin^2\theta d\varphi^2$$

I have used the Schwarzschild radius, R_S, as a replacement for its value: ${2GM}/{c^2}$. In SUiTe, it is impossible for any gravitational source to shrink below its Schwarzschild radius, so the use of R_S is only to simplify the equation.

The right-hand side of the equation above provides us with a great deal of information. Let us imagine a photon of light traveling through a gravitational field. The first term tells us that light is slowed by a gravitational field because the inverse of the square of the Einstein factor is necessarily less than 1. The second term describes the path. Without a gravitational field, light flies in a straight path in three dimensions. Because the Einstein factor is greater than 1, the path will be longer than a straight path would have been; the path of light will be curved. The other two terms have no factor in front of them, which indicates that at the given distance *r* from the source of mass that creates the gravitational field there is a spherical surface where the gravitational field does not vary.

3.3 General Relativity and Special Relativity

The purpose of special relativity is to explain changes to clocks and rulers resulting from motion. The purpose of general relativity is to explain the change to clocks and rulers resulting from gravitational fields. The changes to distance in both special and general relativity are apparent and not real, while the changes to mass and time are real for both. The deformation of space surrounding a source of gravitation cannot be made relative to a location in space that is not similarly deformed. Like gravitation in general relativity, velocity is absolute and not relative in special relativity. Special and general relativity differ in the lack of acceleration in special relativity, and the lack of a direction for the increase in the inertial mass in general relativity. With both the Lorentz factor and the Einstein factor, we are interested in the density of space, and not whether this change in density is due to constant motion or due to gravitational acceleration. It is only when we combine changes in the density of space due to both gravitation and motion, that we will be interested in the source of the change in density.

Relativity is not, as we now realize, the best choice of a name for a theory that has only one relative component, velocity. Even velocity is better thought of as being absolute rather than relative. It was Poincaré who used the name *relativity* to describe the theory that he and Lorentz were working on. The name was then assigned to Einstein's theory about the change to clocks and rulers due to motion. Einstein is believed to have preferred the name *invariance theory*; Maxwell's equations were invariant to the translation of spacetime between two objects that were both in constant motion but moving relative to each other. The measurements of light, electricity, and magnetism are not affected by motion. A laboratory in constant motion will find Maxwell's equations to be true no matter how fast it is moving. General relativity is the extension of this same rule to gravitational fields.

3.4 Simplifying General Relativity

With ten second-order partial differential equations after reductions, general relativity is difficult. At the present time, excluding solutions involving charged bodies, which are relatively simple extensions of the uncharged solutions, there are only five analytical solutions that I am aware of. There is an empty space solution where there are no sources of gravitation. There is the special relativity solution, or Minkowski solution, where there are particles in motion outside of a gravitational field. The Minkowski solution defines spacetime for special relativity. There is the Schwarzschild solution. There is the Kerr solution. Finally, there is the cosmological solution, but this last solution does not appear to be correct. The Kerr solution is important because it combines the Schwarzschild solution with motion, suggesting the possibility that further solutions could be generated by joining general relativity with special relativity.

Using the pressure divided by the density of space, we could now, with the aid of computers, accurately compute the normalized pressure divided by density of any point in space with any number of sources of gravitation. The normalized pressure divided by density of space is a ratio of squared velocities. We can then add the squared velocity ratios of the sources of gravitation together. I don't know whether to credit Einstein or Poincaré, because they both found the same solution: the velocity addition formula. The velocity addition formula makes certain that the addition of two velocities does not exceed the speed of light. By using the velocity addition formula, the Schwarzschild solution can be generalized to join multiple gravitational sources in order to determine the clock and ruler at any given point in space that is subject to multiple gravitational fields in the absence of motion.

What if we wish to understand the path, and changes to mass and time, of an object *moving* in a gravitational field subject to multiple gravitational sources? As explained in the previous paragraph, we can determine the clock and ruler at every point in space subject to multiple gravitational sources. A moving particle carries a clock and a ruler. How can we combine the contraction of space due to gravitation with the compression of space due to motion?

The velocity addition formula provides more than just a clock and a ruler at a point in space subject to multiple gravitational sources. For a moving particle, it also provides direction. If the motion is due only to gravitational attraction, the direction will be toward the nearest location in space that has the slowest clock relative to the clock of the particle. The combination of gravitational sources using the velocity addition formula gives the clock and ruler at any given point that the particle can go toward. If we do this for many points, we can sketch a path that always leads from the particle's location toward the location with the slowest clock in the field.

If the particle is not in free-fall in a gravitational field, it is necessary to consider how the gravitational field affects both its speed and its direction. To determine the path of an already moving particle in a gravitational field with multiple sources, it is necessary to add the vector resulting from special relativity to the vector resulting from adding together multiple gravitational sources and finding the slowest clock. The resulting combined vector indicates the direction of motion of the particle as well as giving the overall change to mass and time for that particle.

In the next section, I will give an example of how it is possible to combine the Schwarzschild solution for a non-spinning, non-moving object with the angular momentum of a spinning object, and how it affects the clock and ruler, and also the direction, of the gravitational force on a small test particle within its gravitational field. This is the Kerr solution arrived at in a different manner than the path used by Kerr. Since many objects are spinning in space and interact with particles traveling near them, the Kerr solution, properly understood, has broad applicability. Repeated application of the velocity addition formula for multiple gravitational sources, spinning or not, and moving relative to each other or not, provides solutions for all combinations of all objects in our universe.

This method does not allow us to consider the evolution of the universe over time, a subject that will be taken up in the next chapter, because the inertial mass is dependent on the density of space and the density of space decreases over time with expansion.

3.5 An Alternative Kerr Solution

Kerr created a solution for a rotating body that had become deformed by its rotation into an oblate spheroid. An oblate spheroid is a sphere that is flattened at the poles. For a rotating body that has become an oblate spheroid, it is useful to use Boyer-Lindquist coordinates. The translation from polar coordinates to the x, y, and z of Cartesian coordinates is: $x = rsin\theta cos\varphi,\ y = rsin\theta sin\varphi,\ z = rcos\theta$. For Boyer-Lindquist coordinates it is:

$$x = \sqrt{r^2 + a^2}\ sin\theta cos\varphi$$
$$y = \sqrt{r^2 + a^2}\ sin\theta sin\varphi$$
$$z = r\ cos\theta$$

The value of r in the above equations is the same value of r that we can use for the Schwarzschild solution, if we are interested in a particle close to the surface of the gravitational source that is spherical. If we like, we can think of r as the value from the center of the oblate spheroid in the direction of the poles. The value of $\sqrt{r^2 + a^2}$ creates a new value greater than *r*, which we will name *b,* and which is a value that considers how gravitation would change the distance at the surface of the equator from the center of a spinning body relative to the distance to the surface at the poles. For Kerr, a is equal to J/mc, where *J* is the angular momentum of a spinning object, and *m* is the mass of that object. We need to choose a value for *r* such that *b* is greater than or equal to the radius to any point on the surface from the center of the oblate spheroid. Here is the Schwarzschild-like solution for a non-spinning oblate spheroid when we are at a distance *b* with no gravitational field:

$$-c^2 d\tau^2 = -c^2 dt^2 + (b^2/b^2)db^2 + b^2 d\theta^2 + b^2 sin^2\theta \; d\varphi^2$$

We want to recover *r,* which is in the equation for *b* and represents the non-spinning radius. We note that a small change in *r*, *dr*, does not match with a small change in *b* because *a* does not change with changes to *r* but is dependent on how fast the object is spinning, which we will assume does not change. We now introduce a new variable *p,* equal to $\sqrt{r^2 + a^2 cos^2\theta}$. In the absence of spin, $a = 0$ and $p = r$. At both poles, *p* will be equal to *b*, but at the equator, *p* will be equal to *r*.

We should think of p as a gravitational correction and not as a distance. For a non-spinning object that is spherical, the strength of the gravitational field will be equal for any particle on a spherical surface at any radius greater than or equal to the radius from the center to the surface of the sphere. We will imagine that when this object spins up, a column of matter between the surface and the center has the same amount of matter, regardless of whether the end of the column at the surface originates at the equator or at the poles. In this case, a particle above the equator will feel the same amount of gravitation as a particle at the same distance above the poles. The value b developed here will describe an equal gravitational surface that is equal to or greater than the equatorial radius and always greater than the polar radius.

The ratio p/b will allow us to scale the changes to the polar radius *r* with respect to the equatorial (always larger) radius *b*. In the polar direction, changes to the radius are the same as they would be for a sphere of radius *r*, so the ratio p/b must be 1, but toward the equator, the changes to the radius need to be reduced because a change in *r* will result in a smaller proportional change in *b*; *a* depends only on rotation and mass, and not on *r*. We now modify the above equation by introducing *dr* and multiplying though by p^2/b^2 for both the radius and latitude because p/b depends on both. We don't multiply the longitude because p/b does not depend on the longitude.

$$-c^2 d\tau^2 = -c^2 dt^2 + p^2/b^2\, dr^2 + p^2 d\theta^2 + b^2 sin^2\theta\, d\varphi^2$$

I will start leaving off the left-hand side in the subsequent equations to save space. It never changes and is always $-c^2 d\tau^2$. In the above equation we haven't done anything to time, dt, which in the Schwarzschild solution is the inverse of the change to radius, *dr*. We will do that now:

$$-(b^2/p^2)c^2 dt^2 + (p^2/b^2)\, dr^2 + p^2 d\theta^2 + b^2 sin^2\theta\, d\varphi^2$$

The radius *b* is constructed from an *r* large enough that $\sqrt{r^2 + a^2}$ is greater than or equal to the equatorial radius at the surface. The radius *b* can therefore be thought of as the radius of a sphere that encloses the spinning oblate spheroid. We will now need to make an assumption that will certainly be incorrect when the equatorial radius is substantially larger than the polar radius; at the radius *b* that we have chosen, any test particle will feel the same gravitational field, even though the equal gravitational spherical surface of the spinning body will be farther away at the poles than at the equator. One way to think of this is that any radial column of matter stretching from the center to the spherical surface at *b* will have the same mass. The matter density from the center to the equator will then be less than from the center to the poles. Spinning will have caused the particles to move away from each other, but will not have caused a redistribution of the particles of the object from the poles toward the center.

We will deliberately exclude the centrifugal force from our calculation of the gravitational force. Its only influence in this model will be to expand the distance between particles. The centrifugal force can always be added back in at the end, subtracting it from gravitation. We all weigh less at the equator of the Earth than at the poles because the centrifugal force is acting on us in the opposite direction from the gravitational force.

Let's now turn the previous equation into a Schwarzschild-like solution for a non-spinning oblate spheroid with an equal gravitational surface at *b*, by making sure that the modifications that need to be passed on to dr^2 will also be passed on to dt^2. The first step is to make adjustments to *b*, treating it as if it was the radius in a Schwarzschild solution, while taking into account the effect that the unchanging value of a will have on changes to b using the weighting of p/b. For time we have:

$$(R_S/b)(b^2/p^2)c^2dt^2$$

And for the radius *r* we have:

$$(b/R_S)\,(p^2/b^2)\,dr^2$$

As we did in the section on the Schwarzschild solution, we have replaced $2Gm/c^2$ by R_S. We can now write out the complete right-hand side of the equation for a Schwarzschild-like solution when there is a gravitational field:

$$-(1 - R_S b/p^2)c^2dt^2 + (1 - R_S b/p^2)^{-1}dr^2 + p^2d\theta^2 + b^2 sin^2\theta\, d\varphi^2$$

Like all Schwarzschild solutions, the multiplicative terms in front of c^2dt^2 and in front of dr^2 are the inverse of each other. I want to make one more small algebraic change:

$$-(1 - R_S b/p^2)c^2dt^2 + (p^2/(p^2 - R_S\, b))dr^2 + p^2d\theta^2 + b^2 sin^2\theta\, d\varphi^2$$

If we replace *p* and *b* with *r,* and then make a sign change, we have the same solution as the Schwarzschild solution for a non-spinning spherical body.

Now we have to add spin. For SUiTe, spin involves motion, which depends on the special theory of relativity. We will imagine that the massive object we are seeking to describe is a superfluid, so that the particles from which the object is made do not interact with each other except through gravitation. We want to know the speed of a particle directly above the equator, which we will assume is at a distance of b from the center. The speed of this particle is ωb, where b is the radius and ω is the angular frequency. We will want the relative speed (relative to light), which is $\omega b/c$. Kerr set a to be the angular momentum divided by the mass and the speed of light. If we use J to be the angular momentum, then a is equal to J/Mc. The angular momentum formula is $J = I\omega = Mr^2\omega$ when M is the mass of an object made of frictionless particles that are spinning together. Setting $r = b$ in the above formula, we have $a = \omega b^2/c$.

The next piece to consider is what happens to the combined compression and contraction of space at the distance b above the equator of the spinning object. This wasn't necessary in the Schwarzschild solution, but now that we are combining the contraction of space due to gravity and the compression of space due to motion, we need to know how changes to longitude affect our measurements. For a particle at the distance b from a spinning object, neither the distance to the center nor the latitude changes due to spin, but the longitude is continually changing. This is where motion (the special theory of relativity) and gravitation (the general theory of relativity) collide.

We can use the velocity addition formula for orthogonal velocities. The formula is $\sqrt{U^2 + V^2 - U^2V^2}$. Both U and V are velocities divided by c. Let's set $U^2 = R_S b/p^2$, the gravitational contraction, and $V^2 = (\omega^2 b^2/c^2)sin^2\theta = (a^2/b^2)sin^2\theta$, the compression divided by density of space due to motion. We realize that we've already taken care of U^2, which was part of the Schwarzschild solution at radius b. We don't care about V^2 because it is orthogonal to the gravitational field and it won't have an added effect on the clocks and rulers at the location we are interested in, but we do care about the interaction effect, the $-U^2V^2$ term:

$$(R_S b/p^2)(a^2/b^2)sin^2\theta = (1/p^2)(1/b^2)(R_S ba^2)sin^2\theta$$

This interaction effect is with the longitudinal term and has to be added to our earlier equation that gave the Schwarzschild solution for a non-spinning body after multiplying by b^2:

$$-(1 - R_S b/p^2)c^2dt^2 + (p^2/(p^2 - R_S\, b))dr^2 + p^2d\theta^2 + (b^2 + (R_S ba^2/p^2)sin^2\theta)\, sin^2\theta\, d\varphi^2$$

The final step is realizing that wherever we see R_S involved, we know that there will be a time effect, so there will be a non-zero value for $dtd\varphi$. As before, when we developed the Schwarzschild solution, the changes are only to the terms involving R_S. To get the $d\varphi$ term, we need to take the square root of $((R_S ba^2/p^2)sin^2\theta)\, sin^2\theta\, d\varphi^2$. To get the dt term, we need the square root of $(R_S b/p^2)c^2dt^2$. Then we multiply the square roots together. That will give us $(1/p^2)R_S bacsin^2\theta dtd\varphi$. Because there is both a $dtd\varphi$ and a $d\varphi dt$, there will be two such terms in our equation. There is still the question of the sign of this term. Because motion increases the length of a second just as gravitation increases the length of a second, this last term has to share the sign of the time term, which is negative. Our final equation is:

$$-(1 - R_S b/p^2)c^2dt^2 + (p^2/(p^2 - R_S\, b))dr^2 + p^2d\theta^2 + (b^2 + (R_S ba^2/p^2)sin^2\theta)\, sin^2\theta\, d\varphi^2 - (2/p^2)R_S bacsin^2\theta dtd\varphi$$

If we were to replace each $R_S b$ with $R_S r$, we would have the Kerr metric. This metric will, like the Kerr metric, be a solution of the equations of the general theory of relativity. What is remarkable is how we found it. Kerr created this equation in a slightly different form more than fifty years ago. We found a way to reach roughly the same conclusion from a very different theory of how the universe is put together.

The value for a depends on the angular momentum, J, which for something made of frictionless particles is $J = I\omega = Mr^2\omega$. If the particles from which an object is made stick together, then the value for J will be less. Recently, some researchers have suggested that collapsed stars may have superfluid interiors. Superfluids were originally discovered by cooling helium to near absolute zero. The atoms of helium slipped past each other without any friction. We will assume that a similar superfluid state can exist for stars with masses greater than 2.3 times that of our Sun. If we use s for superfluid (no stickiness) and set it equal to 1 for superfluidity, and set it to less than 1 if the particles stick together, then we have $J = I\omega = sMr^2\omega$. To turn the above Kerr-like solution into a class of solutions, every a must be replaced by sa:

$$-(1 - R_S b/p^2)c^2dt^2 + (p^2/(p^2 - R_S\, b))dr^2 + p^2d\theta^2 \\ + (b^2 + (R_S bs^2a^2/p^2)sin^2\theta)\, sin^2\theta\; d\varphi^2 \\ - (2/p^2)R_S bsacsin^2\theta dtd\varphi$$

where $b = \sqrt{r^2 + s^2a^2}$ and $p = \sqrt{r^2 + s^2a^2cos^2\theta}$.

Returning to the earlier solution, without the parameter associated with superfluidity, we see the same frame-dragging effect that Kerr saw. Kerr and others interpreted his solution as implying that space could be dragged by a rotating body. A particle or a photon of light that entered into this dragged space would appear to have sped up for those observers looking from the outside. For objects that are very large and spinning extremely fast, space could potentially rip if the space-dragging exceeded the speed of light. In SUiTe, space is not being dragged and cannot be ripped.

Inertial mass has direction in the special theory of relativity. The interaction terms combining the contraction and compression of space must also have direction. The $-U^2V^2$ term in the velocity addition formula for orthogonal velocities has a direction at an angle between the two orthogonal squared velocities, the one due to gravity and the other due to angular momentum. This angle indicates that a particle attracted to a spinning body will feel the gravitational center in advance of the physical center of the spinning body and in the direction of that body's spin. This advance will be greatest for a particle attracted toward the equator of the spinning body. What has been called "frame-dragging" is a displacement of the center of mass in the direction of spin, as felt by a particle attracted to a spinning body.

This new Kerr-like solution opens up problems associated with gravitational fields to possible solutions where objects are spinning or in orbit around each other by calculating the pieces separately and adding them together using the velocity addition formula. Unlike the Kerr solution, this solution requires the existence of an equal gravitational spherical surface at some distance from the center of the source of the gravitational field. As long as that is approximately true, this solution can be combined with others, again using the velocity addition formula, to arrive at solutions for multi-body problems.

Chapter 4

Cosmology

The present theory of cosmology extends general relativity through time, assuming that neither clocks nor rulers change. There are a number of problems with this approach. Why shouldn't clocks and rulers change over time due to the expansion of space? Or at least clocks? Space must have been denser in the past than today. The density of space is responsible for changes to clocks and mass in both special and general relativity, why not in the past? General relativity is only appropriate for any system in which we don't have to concern ourselves with changes to the density of space resulting solely from the expansion of space. We can apply it to systems in the past and it will work perfectly in terms of how objects interacted in the past. The present model of cosmology fails when we seek to understand how the expansion of the universe changes the density of space.

The Hubble constant, because it changes with the expansion of space and measures the repulsion of the particles of space for each other, will help us to understand how the universe evolves. The beginning of the universe is described in the appendix. In this chapter, we will begin at the point where protons, antiprotons, neutrons, and antineutrons can form. The next important milestone in the evolution of our universe is the appearance of dark matter and the disappearance of positrons. After that comes recombination, the moment at which hydrogen atoms were able to form. In previous descriptions of the beginning of our universe, particles of dark matter had been left out. They play a fundamental role in SUiTe.

4.1 In the Beginning

The beginning universe was dominated by the electrostatic force. Space was much denser than it is now, so, by our standards today, seconds were extremely long. With much longer seconds and particles that had far more inertial mass than the same particles have today, the electrostatic force would appear weaker in comparison with today because it had to move much more massive particles than today, but its affect was no different. The energy necessary to excite an atom of hydrogen was less, and the energy given off by hydrogen as it returned to its rest-state was less by the same proportion. This does not mean that the Coulomb force constant was necessarily different, at least not with respect to other force constants, but only that the force and energy associated with the Coulomb interaction, because it is divided by seconds squared and because mass in the numerator is not squared, must have been weaker.

Gravitation began with the first protons and antiprotons. These two particles attract space. The contraction of space around matter results in gravitation. In the very early universe, protons and neutrons were continually forming and disintegrating through contact with antiprotons and antineutrons. The particles that make up space repel each other causing space to expand. As space expanded, the inertial mass of protons and other massive particles decreased.

With an ideal gas, the mass of the particles of the gas does not change. The radiation is proportional to the number of collisions per second. As the gas expands, the number of collisions per second decreases and the temperature decreases. In SUiTe, the expansion of space does not decrease the number of collisions per second because the expansion in terms of an increasing distance between particles of space is apparent but not real. The number of particles of space between the particles of matter has not changed. The decrease in temperature in the early universe is due to the decrease in the inertial mass of the massive particles that are colliding as space expands.

Initially, as the temperature decreased, protons became stable. Later, dark matter was able to form, removing all free positrons and an equal number of electrons from our universe. Particles of dark matter are numerous in comparison to protons, with about 33,000 particles per proton or neutron. Particles of dark matter are unable to radiate and can only reduce their energy through collisions with each other, with photons, and with normal matter.

This results in a gas-like pressure that is opposed to gravitation. As the temperature of the universe dropped to the point where positrons disappeared, dark matter formed everywhere, surrounding protons and neutrons, temporarily preventing them from coming together under the gravitational force.

The process preventing the rapid collapse of giant molecular gas clouds into stars is complicated (see section 6.3). The first stars, using light pressure and stellar wind, were able to expel dark matter from the surrounding gas cloud and tip the balance away from the dark matter pressure that prevents further collapse of hydrogen and helium into additional stars. The initial stars effectively create star nurseries by pushing away dark matter.

The sudden inflation of the universe does not exist in SUiTe. Prior to the formation of dark matter and the subsequent disappearance of positrons, there were about 33,000 electrons and 33,000 positrons for each nucleon. The electrostatic force and radiation dominated the universe. Until dark matter could form, removing huge numbers of electrons and positrons and putting an end to the radiation period, every nucleon in the universe would have been continuously bombarded with radiation and struck by energetic electrons and positrons.

Once dark matter formed, it became another source of repulsion, preventing nucleons from coming together. As explained in the next chapter, at very small distances, matter repels matter with a force that is stronger than gravitation. The homogeneity of the universe began with both electromagnetic radiation, and electrons and positrons striking nucleons. When dark matter formed, positrons and most of the electrons disappeared. But dark matter became the source of matter-to-matter repulsion between dark matter and nucleons which kept the nucleons of the universe apart. These two epochs at the beginning of the universe, one before positrons disappeared and the other after, resulted in the homogeneous distribution of matter that we see today. There was no need to communicate from one side of the universe to the other in order to preserve the homogeneity of the universe and the resultant homogeneity of the cosmic microwave background radiation (CMB). There is no need for an inflation theory.

4.2 Changes to Constants Due to Expansion

Nature measures distance by counting the number of particles of space between two points. Distance measured in this way does not change with expansion. Inertial mass is determined by the number of particles of space that a particle of matter is in contact with. Matter cannot expand as space expands, so inertial mass decreases. Physics time is not an independent quantity but is the repetitive motion of matter. As inertial mass increases, going into the past, time slows and seconds become longer by the same proportion that inertial mass grows.

In SUiTe, there are eight fundamental constants, the gravitational constant, the Coulomb constant (electric permittivity), the elementary charge, the Planck constant, the equatorial area of the protino, matter-to-matter repulsion, the fine structure constant, and the speed of light (See chapter 21). All are presumed to be ultimately dependent on the ratios of the three fundamental forces. The particles exhibiting these forces can only change over time if the ratios of the fundamental particles change.

If there was a process by which the total number of electrostatic charges in the universe decreased, the Coulomb force constant in the past would have been weaker than it is today because, in SUiTe, the rarer a force is, the more powerful it is. The only process in the past that could have been used to reduce the number of the charges in the universe was the possible burning of positrons and electrons at the centers of the early, extremely powerful, quasars. In SUiTe, what we perceive as the annihilation of an electron by a positron is the creation of a particle of dark matter. The true destruction of an electron and a positron would create vastly more energy and would reduce the total number of charged particles in the universe by two.

When a photon is released by an atom, it is the result of the bremsstrahlung of an electron. In the early universe, the acceleration of the electron was slower by $(1+z)^{-2}$ where $(1+z)$ is the redshift plus 1. Mass was greater. The energy produced was the usual energy that we see today for the same process multiplied by $(1+z)^{-1}$. This same process occurs in the transverse Doppler shift of special relativity and the gravitational redshift of general relativity. The energy we measure of a photon that has arrived from the past tells us about the pressure divided by the density of space in the past and its effect on matter (see the two previous chapters).

In order to determine the energy of a photon in the past, if the Coulomb constant and the electron constant have not changed, the only concern is the overall change to mass and time. In SUiTe, the gravitational constant is not separated from the other fundamental constants, but is determined by them (see section 21.1). After the multiplication by two masses, both of which have increased by $(1 + z)$, the gravitational force is changed by the same proportion as the other force constants. All forces and energy in the past decreased by $(1 + z)$. The gravitational force that pushes matter together when compared to the electrostatic force that holds atoms together shows no change going into the past. But when we look at a photon from the past, we see that its energy is less than we would expect by precisely $(1 + z)$.

To determine the changes to constants in SUiTe with expansion, it is necessary to treat distance in the way that nature treats distance: as the number of particles of space between two points. Nature's meter does not change going into the past, but mass and seconds both change by the same proportion. Newtons and joules must both be divided by $(1 + z)$. But because all the constants change in this way, Maxwell's equations remain the same, and special and general relativity remain the same. It is only by comparing the energy of a photon that we receive from the past with the energy of a photon produced by the same process today that we can see the effects of the contraction of space and the resulting increases to the inertial mass of matter in the past.

4.3 The Cosmological Redshift

We will start with the definition of the redshift using the letter *z* for the redshift.

$$z = \lambda_t / \lambda_0 - 1$$

The wavelength is denoted by λ. The subscript *t* refers to a time *t* seconds ago. The subscript *0* is now. If a hydrogen atom emits a photon at t seconds in the past, then it will have a longer wavelength (less energy) than a similar photon emitted by a hydrogen atom 0 seconds in the past. The cosmological redshift is between 0 and a large number which we don't know, but the largest wavelength from the earliest light we have been able to measure suggests a redshift of about 1100. This light comes from the moment when the temperature of the universe had dropped low enough, so that hydrogen atoms could form. There is a gravitational redshift, also between 0 (no gravitational field) and a large number. In this case, if singularities exist, then the redshift could presumably go to infinity, but light would become trapped in a black hole gravitational well before that. The largest numbers for the gravitational redshift that we have been able to measure are not very large at all.

In both cases, from the cosmological past or from a gravitational well, the redshift corresponds to a lengthening of the wavelength of light. For gravitation, using the Schwarzschild solution, the relation is:

$$(1+z)^2 = 1/(1-2GM/rc^2)$$

We can then derive z, the gravitational redshift, from this equation. Is the gravitational redshift different from the cosmological redshift? Not by much. The gravitational redshift increases when traveling along a radial toward the center of the source of gravitation. The cosmological redshift increases along a *time* radial pointing toward the beginning of the universe. To create a cosmological solution similar to the Schwarzschild solution from gravitation, we will need to use a constant which describes the expansion of space. This will be done in section 4.5.

4.4 The Hubble Constant

The Hubble constant is the expansion rate of the universe. Although the rate of expansion depends on time, and it changes over time, the Hubble constant is locally constant in the sense that, at a given moment, we will see the galaxy clusters in the universe moving away from each other at the same rate. Recent measurements have shown that the expansion rate is increasing over time, indicating an accelerating expansion of the universe. We will derive the Hubble constant for a universe with space, but with little matter. In that case, the expansion won't be temperature dependent, nor will it be held back by the attraction of space for matter. We know that this isn't true of the early universe, but, by now, the change in temperature is very small as the universe expands. We can think of this model as an approximation that can be used for the future and much of the past but not for the distant past. The expansion of space will be similar to the expansion of an ideal gas where temperature doesn't change with expansion. We can use Boyle's law:

$$P_0 V_0 = P_t V_t = constant$$

The equation states that the pressure times the volume of space in the present, more expanded universe, must equal the pressure times the volume of space in an earlier, less expanded universe, at time t. It should be noted that the particles of space, which repel each other in SUiTe, have no mass and should, therefore, have no pressure, but the particles of space when compressed in a photon do have light pressure. This is not the same as a gas composed of massive particles, and the behavior of space need not follow from Boltzmann's theory about gases. If we were nevertheless to imagine that Boyle's law was appropriate for empty space, we could potentially describe the behavior of the universe as it expands.

The fundamental constituents from which matter is constructed—the quarks and leptons of SUiTe, so to speak—don't expand as space expands, but the number of particles of space with which they interact decreases. Since, in SUiTe, inertial mass is the interaction of space with matter as matter moves, inertial mass would increase if we could time-travel backward. The increase in inertial mass would be proportionately equal to the increase in pressure. The pressure of space will always increase moving into the past and decrease as we move into the future. We then have:

$$m_t/m_0 = P_t/P_0 = V_0/V_t$$

where m_t/m_0 represents the inertial mass of a particle in the past divided by the same particle today. It is important to note that if we could time-travel to the past, we would not see any difference in masses or in the gravitational constant. Everything would behave precisely as it does now and the fundamental constants would have the same values. The only way of inferring the increase of the inertial mass of matter in the past as compared to today is to measure the wavelength of light from the past using todays clocks. Like the apparent (but not real) contraction of matter that results from using a human devised ruler rather than nature's ruler, the use of clocks—irrelevant from the point of view of a nature which only exists in the present—creates a false view. The contraction of space due to gravitation or the contracted space in the past does not change matter, but can change the measured energy of a photon sent from more contracted space into less contracted space. The simplest way to deal with this problem is to separate entirely the concept of inertial mass, which is dependent on the density of space, from rest mass which, like the electrostatic charge of an object, is independent of any surrounding conditions. Inertial mass and rest mass can only be assumed equal at a given time at non-relativistic speeds and in weak gravitational fields.

Wavelength is related to energy by inverting the wavelength and multiplying by the Planck constant and the speed of light:

$$E = h\,c/\lambda$$

A proportional change to the inertial mass of matter is equal to a proportional change to the associated clock. In the special and general theories of relativity, the increased inertial mass and the slowed clock reduce the energy of the light emitted from either a moving object or an object in a gravitational field, respectively. I suggest that an atom in the past will find its inertial mass increased and its clock slowed, both by the same proportion.

$$m_t/m_0 = E_0/E_t = (hc/\lambda_0)\,(hc/\lambda_t)^{-1} = \lambda_t/\lambda_0$$

The second and third terms represent the energy of a present photon divided by the energy of a past photon. Note that the mass ratio is the inverse of the energy ratio. We could use the energy of a photon emitted by hydrogen today, divided by the energy of a photon from hydrogen emitted from a distant star, which would be light emitted in the past.

In SUiTe, the redshifts of the special theory of relativity, the general theory of relativity, and cosmology are all the result of the inertial mass and clock change of the emitting particle. The redshift is not caused by a stretching of space because space is made of particles and cannot stretch in SUiTe, assuming that nature's ruler is based on the number of particles. With changes in inertial mass come changes in the energy of the emitted photons, as described above. Putting this all together, we have

$$m_t/m_0 = P_t/P_0 = V_0/V_t = \lambda_t/\lambda_0 = 1 + z$$

where z is the redshift. The last two terms above are the definition for the redshift. With some algebra, we see that

$$z/(1+z) = (V_0 - V_t)/V_0 = (m_t - m_0)/m_t$$

The middle term above is the percentage change in volume between now and t seconds earlier. If we set t in the near past for a very small change in the volume, and divide by seconds, we will have the percentage rate of change of the universe, the Hubble constant. In this model, where the amount of matter is considered to be too small to slow the expansion of space, the Hubble constant will not only be locally constant, but will always be equal to the inverse age of the universe in seconds. If we know the redshift and the Hubble constant, we can determine how long ago the light was emitted by using this equation:

$$z/(1+z) = H_t t$$

We can see that for nearby objects, where z is much less than 1, the left-hand side is approximately z. Multiplying both sides by c, we have the recessional velocity associated with the expansion of space that is approximately equal to $H_0 d$, where H_0 is the Hubble constant as we measure it now, and d is the distance to the object that is receding:

$$zc = H_0 d$$

This is the equation that Hubble found. Looking back in time, 14 billion years ago, the inertial mass of an atom emitting light will have increased, and the redshift will therefore also have increased. z will then be large, much greater than 1, and the left-hand side of

$$z/(1+z) = H_0 t$$

will be approximately equal to 1. In that case, the Hubble constant will be the inverse age of the universe in seconds.

In a universe with no matter, the Hubble constant will always be the inverse age of the universe in seconds, but because matter slows the expansion of space by attracting the particles of space, the Hubble constant will always be less than the inverse age of the universe, suggesting to us that the universe is older than it actually is. The Hubble constant will increase over time as the increase in the inertial mass of matter due to the contraction of space decreases, ultimately reaching the point where it is equal to the inverse age of the universe in seconds. This model requires an accelerating expansion of the universe because matter interacts with fewer particles of space as space expands, and it is matter that is slowing the expansion of space.

The measured Hubble constant of today is already close to the inverse age of the universe, judging by the age of the oldest stars, suggesting that we have entered an expansion-dominated phase of the universe and have left a matter-dominated phase. This expansion corresponds to what is called “dark energy.” Strictly speaking, it is not energy because space has no mass, and dark energy does not contribute gravitationally.

4.5 A Schwarzschild Solution for Cosmology

We will do as we did before with the Schwarzschild solution; look for an acceleration that we can use to create a factor similar to the Lorentz factor. First we need an equation for the Hubble constant in terms of $1 + z$:

$$1/(1+z) = 1 - H_0 t$$

$$(1/(1+z))^2 = 1 - 2H_0t + H_0^2t^2$$

We need an acceleration and a distance, as before. The distance in this case will have to be $d = ct$. The acceleration will have to be H_0c. Rewriting the equation in terms of a velocity squared divided by the speed of light squared, we have:

$$(1/(1+z))^2 = 1 - 2H_0cd/c^2 + H_0^2d^2/c^2$$

Unless $t = {}^{d}/_{c}$ is close to the age of the universe, $H_0^2d^2/c^2$ is going to be very small because the Hubble constant is very small, about $2.3 \cdot 10^{-18}$ inverse seconds today. When the Hubble constant is squared, except for a very large *t,* it won't be worth calculating. If we use only $1 - 2H_0cd/c^2$, we can generate a Schwarzschild-like solution for clocks and rulers going back in time. This will be similar to the Schwarzschild solution in general relativity because we have

$$(1/(1+z))^2 = (1 - 2GM/rc^2) \cong (1 - 2H_0cd/c^2)$$

We can also simplify $1 - 2H_0cd/c^2$ (remembering that $d = ct$) to $1 - 2H_0t$ to arrive at a Schwarzschild-like solution.

$$c^2d\tau^2 = (1 - 2H_0t)c^2dt^2 - (1 - 2H_0t)^{-1}dr^2 - r^2d\theta^2 - r^2sin^2\theta d\varphi^2$$

Once again, we are using polar coordinates to describe a spherical surface surrounding something, but this spherical surface is an equal-time surface, and it does not surround the creation of the universe, but instead surrounds us at the present time. In whichever direction we look, at a given time in the past, clocks were slower and the energy of an emitted photon was less than they are today. In the Schwarzschild solution, the radius points at the source of a gravitational field. In this solution, which we could call *the Hubble solution*, the radius points at us.

4.6 Two Ways to Measure Distance

Nature is quantum. For nature, distance is the number of particles between two points. The rest of nature is also quantized. For example, there is a quantum of charge. There are no partially charged objects that have ever been observed.

Finally there is time. Here nature is not very helpful. There is biological time; something must be born before it can die, and there is physics time: the repetitive motion of matter, a pendulum for example. Nature prefers biological time. In order to predict what will happen in the future, we need to use physics time. One second is 9,192,631,770 transitions of the outermost electron of cesium 133. The meter is 299792458^{-1} of the distance light travels in one second.

It is not surprising that cosmology will look different with nature's ruler. A ruler that depends on a specific number of particles of space will not be able to match a ruler that depends on a second, since seconds vary due to velocity, gravitation, and the changing density of space. The expansion of space does not change the number of particles of space and doesn't change nature's ruler, but it does change seconds. Going into the past, using our physics ruler, the universe is shrinking, and going into the future, it is expanding. With nature's ruler, the size of the universe does not change. Inertial mass and seconds change.

With nature's ruler, inflation theory could never have worked. The speed of light depends on the number of interactions between the two particles from which space is made. If the ratio of these particles in space doesn't change, the speed does not change. Using nature's ruler, the speed of light does not change going into the past. Using our ruler, it does. The distance of our meter in terms of the number of particles of space that fit into it increases. There are a limited number of particles of space, so eventually the entire universe is less than a meter in radius. The universe ultimately shrinks into a singularity. Singularities are non-physical.

But with a universe that is very small, light energy can distribute itself evenly throughout the universe. If the universe were to inflate at that moment faster than the speed of light, the equal distribution of mass-energy could be maintained until a time when inflation ended, and the universe was larger. The universe today would have roughly equal amounts of mass-energy in all directions, in spite of the after-inflation expansion. This solution comes with many problems. Two of them are: why did inflation start, and why did it stop.

The other way to look at this is that, until the time when dark matter formed and positrons disappeared, normal matter was distributed equally because gravity was far too weak in comparison to the Coulomb force to interfere with the combination of light energy and the rapid motion of electrons and positrons. When dark matter formed, the radiation dominated period of the universe ended. After that time, the primary difficulty in creating giant molecular clouds that can later collapse, creating the first galaxies, is getting rid of heat. The formation of dark matter not only generates a great deal of radiation in the form of gamma rays, but it also greatly decreases the loss of kinetic energy due to radiation, because dark matter can only lose kinetic energy via collisions. The mixed plasma of normal and dark matter remained too hot to collapse under gravitation until after recombination. Until then, both normal and dark matter would have been evenly mixed.

There is no reason why we need to avoid nature's ruler and use only our own. We may not like the idea that the universe is the same size now as it has always been, but it doesn't change what we observe. At a given moment in time, the two can be made the same. The meter today corresponds to a bit more than 10^{18} w-dual particles. Our meter in the distant past would have had far more particles in it, and our meter in the future will have slightly less, if we continue to use the distance light travels in one of our seconds. There are some advantages for the way we do physics in using our choice for seconds to determine distance rather than nature's choice of using a fixed number of particles for distance. But these advantages cause problems when we look at the way the universe evolves over time. It isn't that there is a right way and a wrong way to define distance, but that our choice of how we choose to measure distance will give incorrect answers if applied to the wrong problems.

4.7 Cosmic Distance Ladder

By first measuring the distance to nearby stars using parallax, and then choosing standard candles—certain stars or other objects that have the same brightness—it is possible to determine the distance to distant objects. Another goal of creating a cosmic distance ladder is to arrive at an accurate value of the Hubble constant. In the SUiTe cosmological model, the Hubble constant takes the place of the gravitational constant in the corresponding Schwarzschild solution for cosmology (see section 4.5).

Both the WMAP and Planck telescopes provided maps of the tiny variations in temperature of the CMB radiation. This should be a picture of the universe at the time of recombination, when hydrogen and helium atoms were able to form. Using the ΛCDM model of cosmology, it is possible to obtain an accurate estimate of the Hubble constant from these maps. Unlike the distance ladder, this measurement is model dependent.

Λ (lambda) is a measure of the dark energy component of the total energy of the universe on the assumption that the total energy is the amount necessary to flatten the four-dimensional universe of general relativity into the three dimensions that we observe. In SUiTe, no energy is required for expansion because co-moving accumulations of matter maintain the same distance from each other. However, when we measure distance, we perceive that they are moving away from each other. This is the result of having two different ways of measuring distance as described in the previous section. For nature, distance is the number of particles between two points. This does not change with expansion. Because there is no force pushing galaxy clusters away from each other, there is no energy involved and there is no need for something that describes dark energy.

Although SUiTe does not deliberately get rid of Λ because of its problems, it can be noted that the two largest fine tuning problems in physics are associated with dark energy. Why should nature continue to add precisely the necessary amount of space, as space expands, in order to flatten the four-dimensional universe into three? What is the source of the energy that is almost, but not quite, equal to the quantum vacuum energy—an energy that is 120 magnitudes larger than the dark energy of cosmology—such that adding the quantum vacuum energy to this other undefined energy of space results in the tiny residual energy that is dark energy?

The other difficulty with a measurement of the Hubble constant based on the early universe CMB map, or on baryon acoustic oscillations (BAO) that depend upon the same CMB map, is that the temperature differences that they are attempting to measure arise during the short period of rapid inflation of the universe in the first seconds after the Big Bang. This inflationary period does not exist in SUiTe. That the values found for the Hubble constant from the CMB analysis or from BAO are very close to each other is not surprising since both depend on a specific theory of how the early temperature differences arose in the CMB. That this value is not distant from the value found from the cosmic distance ladder approach is interesting but not compelling. The distance ladder is the only theory-free approach to determining distance.

For the distance ladder, when using standard candles, it is necessary to compare the apparent magnitude measured by a telescope on Earth to the absolute magnitude at a standard distance from a standard candle. If the standard candle is a star, the magnitude of the star that would be measured at a distance of 10 parsecs is used as the absolute magnitude. If both the absolute and apparent magnitudes are known, it is a simple matter of geometry to determine the distance. In the case of a type Ia supernova, the absolute magnitude is believed to be about -19.5.

There are a number of potential problems with the apparent magnitude. The telescope can measure the brightness at different light frequencies, but the brightness is affected not only by distance, but also by molecular clouds and dust in the interstellar medium. It is also possible that the distant supernova has been weakly gravitationally lensed. SUiTe raises an additional difficulty; the light from the distant supernova may have been affected by passage through molecular clouds containing dark matter. This may be a difficult problem. Different frequencies of light respond differently when passed through dust or through clouds of gases. Using a variety of frequencies, it is possible to adjust for the interstellar extinction that would make a star appear to be further away than it is. But it is possible that dark matter will affect all frequencies equally, except for very low frequencies at the microwave or radio-wave level. More distant objects, with their light having passed through more dark matter, may be dimmer than they should be at all frequencies.

In the absence of any light extinction, the relationship between the apparent magnitude and the absolute magnitude is straightforward in SUiTe. The magnitude is a measure of the luminosity. The luminosity is in watts—joules per second. The energy of a photon produced by the same process in the past is reduced by $1 + z$ as compared to the same process today, where $1 + z$ is the redshift plus 1. Seconds are lengthened by the same amount. This results in a very simple formula for the luminosity of the past.

$$L_t = L_0 \div (1 + z_t)^2$$

When Einstein developed the theory of general relativity, he saw that light would be bent when traveling through a gravitational field. In 1914, a German eclipse expedition went to Russia to look for this effect, but they were arrested at the outbreak of the war. The lack of a measurement in 1914 turned out to be fortunate for Einstein. He had seen the spatial effect of passing light through a gravitational field but not the time effect. The time effect is exactly equal to the spatial effect. The bending of light is twice what Einstein had predicted in 1914. The corrected version of general relativity was published in 1915. There should be both a space and a time effect in cosmology as well.

The initial $1 + z$ is the space effect and can be used to determine distance as Hubble did. The time effect in general relativity is not only a doubling of the space effect; the time effect exists on its own in cosmology just as the Shapiro effect does for general relativity, the time delay of signals sent through a gravitational field. Where can we see the time effect by itself in cosmology?

Type Ia supernovas have a specific light curve over the time period that the explosion occurs. Two adjustments must be made in order for Type Ia supernovas to be used as standard candles:

(1) the maximum absolute brightness differs slightly between different types of Ia supernovas, and (2) the light curve may be lengthened or shortened. The lengthening of the light curve for distant supernovas is $1 + z$ times the expected length of the light curve. This is referred to as the supernova stretch factor. This is the time effect of cosmology.

The redshift can be easily determined by measuring the wavelength of a photon produced by a known process in the past and comparing it with the wavelength of the same photon produced today. Using the formula above, in the absence of light extinction, both the apparent magnitude and the absolute magnitude can be determined. In the case of a standard candle where the absolute magnitude is known, it is possible to determine the level of light extinction by comparing the measured apparent magnitude with what it should be.

Hubble Constant

Determining the redshift is easy. SUiTe gives a formula for determining the Hubble constant from both the redshift and the distance. I have described this in section 4.5. The formula is:

$$(1/(1+z))^2 = (1 - H_0 d/c)^2 = 1 - 2H_0 cd/c^2 + H_0^2 d^2/c^2$$

This can be simplified to

$$(z/(1+z)) = H_0 d/c \cong cz = H_0 d$$

for $z \ll 1$. This is Hubble's law, but it was found here by assuming that there was a general-relativistic cosmological solution similar to the Schwarzschild solution. For this solution, we need to assume that going into the past is similar to going into a gravitational well.

The one potential difficulty with the cosmic distance ladder, if dark matter plays a role in light extinction, is that the further away a star is, the more likely it will be that dark matter has dimmed the star. This may result in estimates of the Hubble constant, even within the local supergroup, that are too high.

Chapter 5

Matter-to-Matter Repulsion

Matter-to-matter repulsion is required by SUiTe via symmetry. The particles of space repel each other, resulting in the expansion of the universe. The particles of space are attracted to matter, and matter is attracted to space, resulting in gravitational attraction. Matter therefore should repel matter. Matter-to-matter repulsion must be stronger than gravity. At the same time, it cannot be strong enough to prevent nuclei from forming. Finally, because matter-to-matter repulsion is similar to the repulsion between particles of space, it will be heavily screened by space. It will be a short-range force.

Matter-to-matter repulsion is one of the eight fundamental constants of SUiTe. For that reason, its value can be found from the equation linking the constants in section 21.1. The symbol k_R will be used for the repulsion constant. As explained in section 21.1, k_R has units of m^2/s^1. In order to use it as a force similar to the Coulomb force, though not an inverse-square-law force, it is necessary to multiply k_R by the speed of light and a mass.

The difficulty in using the repulsion force is screening by space. It appears that the inverse fourth power of the distance between two sources of mass works best. It allows neutrinos to go through the Earth with little deflection. A slightly different form of the inverse fourth power helps to prevent small neutron stars from collapsing into black holes. This also prevents the nucleus of lead 208 from adding additional neutrons using only gravitation.

5.1 Repulsion Scattering

In Rutherford scattering, the electrostatically positive gold nucleus repels the positively charged alpha particle using the Coulomb force. Matter is able to repel matter in the absence of any electrical charge, including those due to weak residual charges such as London forces. In the absence of matter-to-matter repulsion, the scattering cross section for the neutrino and a nucleus would be based on the radius of the nucleus, added to that of the neutrino. The neutrino is a cylinder with a radius of about 0.3 femtometers. This area of the neutrino is the same as that of a similar cylinder that can enclose an electron. The equatorial area of the spherical protino from section 21.1 can be used to determine the radius of the neutrino.

If there was no repulsion, with an average particle density in the Earth's crust of $8 \cdot 10^{28} \cdot m^{-3}$ excluding electrons (density of 2,830 kg per meter cubed, average nuclide of 21.14 nucleons), and an average radius of 3.32 femtometers per nucleus, the scattering cross section for the elastic scattering of the neutrino and the nucleus when both are treated as spheres would be $4.1 \cdot 10^{-29} m^2$. A neutrino would be likely to scatter off a nucleus every 3.3 meters. It would not make it through the Earth.

The formula for determining how close the alpha particle is able to come to the gold nucleus in Rutherford scattering is:

$$d_{min} = 2\frac{k_C q_1 q_2}{m_\alpha v^2}$$

where the Coulomb force is divided by the kinetic energy of the non-relativistic alpha particle.

The repulsion force is more complicated. Without the added effect of screening, the force would spread out like photons streaming from the Sun, an inverse-square-law force. If we were to imagine a cylindrical column stretching out from the center of the sun to a detector, where the area of the detector is equal to the area of the cylinder, at a distance not far from the surface of the sun, the number of photons reaching the detector would be close to the number of photons leaving the cylinder. As the distance between the detector and the surface of the Sun grows, it is necessary to apply the inverse-square-law. For repulsion, it is necessary to determine how many nucleons are in the cylindrical column N_c that stretches from the center to the surface of the nucleus. It is also necessary to have the relative area A_v/A_n, the area of a neutrino divided by the area of a nucleon. The area of the column is the area of a nucleon. Multiplying N_c by the ratio of the areas is necessary because the neutrino is functioning as the detector of the matter-to-matter repulsion from the nucleus, and the neutrino-detector has a smaller area than the column of nucleons whose force it is detecting. The area of the neutrino is about $2.9 \cdot 10^{-31}m^2$, and the area of a nucleon is about $2.27 \cdot 10^{-30}m^2$.

Repulsion is based upon treating each nucleon in the nucleus like the Sun, but sending out forcelines in every direction in place of photons. The matter-to-matter forcelines that interact with an exterior particle, such as a neutrino, are a small subset of all of the forcelines exiting the nucleus, just as a photon detector placed above the surface of the Sun would detect only a small proportion of the total number of photons emitted by the Sun.

The column number estimates the number of nucleons that can be found in a cylinder one nucleon wide linking the center of a nucleus or larger object with the surface. For nuclei up to calcium 40, the number of shells is 2 in SUiTe (see section 18.2). The radius of a nucleus is about $\sqrt[3]{A} \cdot R_0$ femtometers where A is the number of nucleons, and R_0 varies between 1 and 1.25 depending on the type of nucleus. 1.2 will be used here for an average nucleus of 21.14 nucleons in the Earth's crust. The average radius is then $r = 3.32$ femtometers. We can also imagine a numerical radius in the case that the volume is equal to the number of nucleons. This will be $r_N = \sqrt[3]{A \cdot 3 \div 4\pi}$. Dividing the numerical radius, after multiplying by femtometers, by the actual radius, and then multiplying by the number of shells, tells us how many nucleons are in the imagined cylinder. N_c reduces the repulsion from the nucleus felt by the external particle because only those nucleons in the column contribute to repulsion, and there is some empty space between nucleons in the column.

$$N_c = 2 \cdot {}^{r_N}/_{r} = 1.033$$

In SUiTe, the neutrino is in the shape of a cylinder twice as long as it is wide whose long axis is parallel with its direction of motion. The radius associated with the area of the cylinder is about 0.3 fm and the area we will use is $2.9 \cdot 10^{-31} m^2$ (section 21.1). The equation for determining the minimum distance between the neutrino and the nucleus is:

$$d_{min} = 2\frac{k_R c \cdot N_c (A_\nu / A_N) m_N \left(d_N^4 / d_{min}^4\right)}{m_\nu v^2}$$

As can be seen from the equation, it is necessary to choose d_{min} according to the kinetic energy of the neutrino because nothing else in the equation changes. The subscript *N* refers to a nucleon. $k_R c \cong 4 \cdot 10^{-7} \cdot m^3 s^{-2}$, $m_N \cong 938.92\, MeV$, $d_N \cong 1.7\, fm$, $m_\nu \cong 0.05\, eV$, and $A_\nu / A_N \cong 0.128$. The value for m_ν is unknown, but recent measurements from the Katrin consortium place it under 0.8 eV. In SUiTe, it is a Majorana particle which places a lower limit of about 0.015 eV. There is also the Planck collaboration which suggests that it is less than 0.09 eV. The value chosen for the mass of the neutrino, as long as it is within a magnitude of the correct value, is not important, but the form of the equation is.

The term d_N^4 / d_{min}^4 in the numerator reflects the screening of space and turns the repulsion force into a short range force. Without this term, repulsion would be a long-range force similar to gravity or the Coulomb force. At very short distances, on the order of a femtometer, the screening by space will begin to decrease and the repulsion will subsequently increase. This will require a modification of d_N^4 / d_{min}^4.

Note that the neutrino scattering cross section is inversely proportional to d_{min} rather than directly proportional because it is not possible to measure the neutrino directly, but it can only be measured by observing a change to a nucleus or an electron that is struck by a neutrino. The larger the minimum distance due to repulsion, the less likely that the neutrino will strike a nucleus. The measured scattering cross section will then increase as the minimum distance due to repulsion decreases. The exponent column in the following table has been added to indicate the possible need for an adjustment to the term used to calculate screening.

E_ν	d_{min}	v_ν	m_ν	$exponent$
100 MeV	$82.86\,fm$	63,246	0.05	4
1 GeV	$51.91\,fm$	200,000	0.05	4
10 GeV	$33.39\,fm$	632,456	0.05	4
100 GeV	$21.06\,fm$	$2.000 \cdot 10^6$	0.05	4
400 GeV	$15.96\,fm$	$4.000 \cdot 10^6$	0.05	4
800 GeV	$13.90\,fm$	$5.657 \cdot 10^6$	0.05	4
1 TeV	$13.29\,fm$	$6.325 \cdot 10^6$	0.05	4
10 TeV	$8.38\,fm$	$2.000 \cdot 10^7$	0.05	4
100 TeV	$5.27\,fm$	$6.325 \cdot 10^7$	0.051	4?
1 PeV	$3.15\,fm$	$2.000 \cdot 10^8$	0.067	4?

In many of the processes in which the cross section can be measured, the cross section divided by the neutrino energy is nearly constant for energies from 2 *GeV* to 350 *GeV*. (For example, see Table 51.2, *Neutrino Cross Section Measurements*, P.A. Zyla et al. (Particle Data Group), Prog. Theor. Phys. 2020, 083C01 (2020)). The equation converting the minimum distance into the cross section for charged current scattering when an antimuon neutrino interacts with a nucleon to create a positive muon and various output particles is:

$$\sigma_\nu/E_\nu = d_{min}^{-1} \cdot (2\pi)^{\log(GeV)} \cdot 1.77 \cdot 10^{-50} cm^3 \div GeV$$

σ_ν/E_ν has units of cm^2/GeV. d_{min} is the distance from the center of the nucleus that repels the neutrino in centimeters. *GeV* represents the energy of the neutrino in *GeV*. $(2\pi)^{\log(GeV)}$ is unitless. The value $1.77 \cdot 10^{-50} cm^3$ in the above equation should depend on the type of neutrino interaction and the type of neutrino. In SUiTe, the muon neutrino is more massive than the electron neutrino, perhaps by as much as three times. The area is also larger. This should result in little or no change to the minimum between the neutrino and the nucleus, and any change to the value $1.77 \cdot 10^{-50} cm^3$ may be small. Nevertheless, $1.77 \cdot 10^{-50} cm^3$ is a proportionality constant. It is the form of the above equation and the equation for the minimum distance which is important.

For the interaction of a muon neutrino with a nucleus of the same size for the antimuon neutrino, the proportionality constant would be $3.54 \cdot 10^{-50} m^3$. I should add that N_c does not vary much with small nuclei, and it is the fifth root of N_c which is used to determine the minimum distance. It is important to note that a number of experiments with different sized nuclei have been done. The results don't suggest a dependence on the size of the nucleus.

At 1 *PeV*, the minimum distance will be close to or less than the radius of the target nucleus. The likelihood of reflection or of deflection of the neutrino at a wide angle is greatly increased. It is believed that the Earth is opaque to very high energy neutrinos. For lower energies, less than 1 *TeV*, those neutrinos that pass close to the nucleus will be slightly deflected, but because the minimum distance is small compared to the radius of an atom, most neutrinos will pass through atoms with little or no effect. What deflections occur are likely to be minor ones randomly distributed around a straight path through the Earth. The situation is different for dark matter.

Dark Matter Scattering

The equation is the same, but two values are now different. $m_{dm} \cong 0.162\ MeV$ and $A_{dm}/A_N \cong 0.363$. Velocities will also be much lower. The most likely velocity for dark matter particles hitting the Earth should be greater than the solar escape velocity at Earth orbit, 42 km/s, but may range higher, perhaps as high as the Sun escape velocity of 618 km/s.

$$d_{min} = 2\frac{k_R c \cdot N_c (A_{dm}/A_N) m_N \left(d_N^4/d_{min}^4\right)}{m_{dm} v^2}$$

E_{dm}	d_{min}	v_{dm}	m_{dm}	*exponent*
81 keV	$422\ fm$	1	$162\ keV$	4
8.1 MeV	$168\ fm$	10	$162\ keV$	4
143 PeV	$5.97\ fm$	42,000	$162\ keV$	4?
3,240 PeV	$3.20\ fm$	200,000	$162\ keV$	4?
30,936 PeV	$2.04\ fm$	618,000	$162\ keV$	4?

Locating dark matter may be similar to locating a neutrino; it will only be seen through its effect on another particle that it has bounced off of or by the dark matter particle breaking into its parts and releasing a positron. In that case, like the neutrino, the minimum distance that a dark matter particle can achieve with a nucleus will be inversely related to the scattering cross section that we can measure.

There are certain Feynman diagrams, associated with the electron in the hydrogen atom, that result in infinities. These infinities are resolved by renormalization. These same diagrams, in SUiTe, result from the interaction of a particle of dark matter with the electron. For that to happen, the density of dark matter in the atom must be nearly constant. Both the value of the repulsion force, $k_R c$, and the minimum distance taken from the above equation must be such that the repulsion force is neither too weak or too strong, and the repulsion distance between the particle of dark matter and the nucleus is not too large. A slowly moving particle of dark matter cannot be pushed so far from the nucleus that no dark matter particles are between the electron and the proton in hydrogen.

If dark matter was like the neutrino, able to travel very close to the nucleus and not feel the repulsion force, dark matter would travel through planets without stopping. The above table shows that dark matter will be widely deflected. Multiple deflections in the crust will slow dark matter. At that point, it will diffuse into the top layer of the crust. Dark matter has little mass and is responsive to light pressure. The rapidly rising temperature of the Earth moving inward from the surface will push dark matter toward the surface. The molten core of the Earth will have little dark matter. Most will be at the surface. The build-up of dark matter at the surface of the Sun will be explained in the next chapter in the section on coronal heating (6.1).

5.2 Adding Neutrons to Lead with Gravity

In chapter 18, *Nuclear Structure and Magic Numbers*, lead 208 is described as having its outer shell full of neutrons and therefore unable to have a stable isotope at lead 210. The reason given was the lack of a place where an additional neutron could share the electrostatic force with a proton in the shell below it. That would not prevent a neutron from being added on top of another neutron using gravity unless there is a repulsive force that exceeds the gravitational force. Repulsion is not a central force. It will be necessary to divide $k_R c$ by meters squared to provide a force. At short distances, between an outer nucleon of a lead nucleus and a neutron attracted by gravity to the lead nucleus, this provides a reasonable result, but the greater the minimum distance, the less useful the equation.

$$\frac{k_R c \cdot N_c m_N \left(d_N^4 / d_{min}^4\right)}{m^2} = \frac{G M_{Pb} m_n}{r^2}$$

$$r = d_{min} + R_{Pb}$$

There is an additional change from the earlier equation describing repulsive scattering. There is no need to adjust the column area because, for both the external neutron and the column, the area corresponds to the equatorial area of a neutron.

The mass of the lead 208 nucleus is about $3.48 \cdot 10^{-25} kg$. The mass of a nucleon is about $1.674 \cdot 10^{-27} kg$ which is slightly less than that of a neutron. The radius of a lead nucleus is determined by using $\sqrt[3]{A} \cdot R_0\, fm$ with $R_0 = 1$. N_c uses the number of shells, 5 for lead (see section 18.2).

$$N_c = 5 \cdot {}^{r_N}\!/_{r_c} = 5 \cdot \sqrt[3]{208 \cdot 3 \div 4\pi} \Big/ \sqrt[3]{208} \cdot 1 = 3.1$$

The gravitational force from a lead nucleus equals the repulsion force at a distance of about 91.5 picometers. Neutrons will have to have more than gravitation in order to join a lead nucleus. What is the velocity at which a neutron can penetrate the repulsion of the lead nucleus and reach its surface? Using the formula for dark matter after adjustment for the neutron, it is $1{,}230\, m/s$. Relatively slow neutrons can reach the surface of lead and be incorporated, providing that there is a space for them. If there isn't, the moment the neutron slows down, it will be pushed away. Clearly, repulsion is not a force that will be strong enough to have a noticeable effect on the construction of nuclei nor on the forces that hold nuclei together.

5.3 Black Holes Can't Collapse

If black holes are singularities, then they cannot exist in SUiTe. The equation in the previous section which made it possible to determine the repulsion force indirectly by measuring the gravitational force at the equilibrium distance between the two forces was not complete as it did not include what I named in chapter 3 as the Einstein factor. It is important to include the Einstein factor for a black hole, because it is not only matter which is being compressed in a black hole but space as well. The Einstein factor measures the contraction of space and its effect on matter. Time slows, and mass increases.

$$\gamma_E = \frac{1}{\sqrt{1 - {2GM}/{rc^2}}}$$

This raises the question of whether compressing matter in a black hole can increase its mass above the mass of the particles from which the black hole was constructed. It cannot. Although the contraction of space does change the inertial mass, it does not change the gravitational mass. Compressed matter is no more attractive to space than uncompressed matter. The number of attractive forcelines of matter that is the cause of the contraction of space around a massive object cannot increase unless the number of massive particles increases. The mass of a black hole remains the mass of everything that went into it, but the inertial mass of the individual particles reflects the difficulty of motion in contracted space. Inertial mass and gravitational mass are not equal in strong gravitational fields.

The previous equation was able to find the repulsion force indirectly by first determining the distance at which the gravitational energy would be unable to bring the neutron closer to the lead nucleus. The question with a black hole is how the repulsion force pushes neutrons apart within the neutron star. This leads to a simpler equation where the sum of the forces between neutrons is described by the ratio of the minimum distance to the actual distance. This ratio is once again to the fourth power and multiplied by the number of neutrons in the column stretching from the center to the surface of the black hole. The previous equation, which did not include the Einstein factor, cannot be used because of the importance of space in resisting gravitational contraction.

The following equation determines the ratio of the distance between neutrons which are touching and neutrons kept apart by both matter-to-matter repulsion and by the repulsion of the particles of space for each other. In order to compare the repulsion force with the gravitational force, it is necessary to divide by square meters because the repulsion force is not an inverse-square-law force, but is a short-range force, really a pressure. This equation only functions at short distances between nucleons.

$$\frac{k_R c \cdot N_c m_n \left(d_n^4 / d_{min}^4\right) \gamma_E}{m^2} \cong \frac{G M_{bh} m_n}{r_{bh}^2}$$

$$r_{bh} = N_c \sqrt[3]{4\pi \div 3} \cdot (d_{min}/d_n) \cdot 10^{-15}\, fm$$

For pulsars, the attraction between shells results in the nucleons of the shells touching each other. It is only after the protons in the shells combine with the electrons that were initially pushed to the surface of the pulsar that the equations above become necessary. In SUiTe, all massive bodies larger than pulsars, including black holes, are made up of neutrons. I will use the terms neutron stars and black holes interchangeably. After reaching the maximum size of a pulsar, the following neutron star initially increases in radius, exceeding the radius of a pulsar, before decreasing asymptotically to the Schwarzschild radius.

I will assume that I know the mass of the neutron star. I will also assume that there is a maximum density for pulsars of $4 \cdot 10^{17} kg \cdot m^{-3}$. I will be able to show that the repulsion equation above prevents this maximum density from ever being violated for objects of any size.

For a density of $4 \cdot 10^{17} kg \cdot m^{-3}$, the radius $r \cong \sqrt[3]{M \div m_n}$ femtometers, where m_n is the mass of a neutron. The numerical radius in terms of neutrons is $N_c = \sqrt[3]{M \div m_n} \cdot \sqrt[3]{3 \div 4\pi}$. For pulsars, r will be based on the density of $4 \cdot 10^{17} kg \cdot m^{-3}$, the radius will increase suddenly when shells of equal numbers of protons and neutrons can no longer be maintained, and the density no longer has a fixed value. At that point, the electrons at the surface are forced into the protons by gravitation, the shells combining protons and neutrons in equal numbers disappear, and the electrostatic attraction between shells is no longer able to assist gravitation in condensing the pulsar. The repulsion force is then able to dominate. The value of the density can no longer be assumed, and it is instead necessary to use N_c to find the radius using the equation above for r_{bh}.

The packing efficiency of a pulsar depends strongly on the diameter chosen for the nucleon. The percentage of space that is filled by the nucleons in the sphere of a pulsar at a density of $4 \cdot 10^{17} kg \cdot m^{-3}$ is close to the maximum percentage for the random close packing of spheres (63.4%).

The odd thing about the Schwarzschild radius is that it increases linearly with the mass. Yet, the gravitational force increases with the inverse-square of the radius. The Schwarzschild radius expands much too fast with the result that the mass density inside the radius is falling rather than rising. This is the result of combining the contraction of space due to gravitation with the compression of matter due to gravitation. I will assume that neutrons cannot be compressed beyond the point where they touch each other. The concentration of both matter and of space results in the very large Schwarzschild radii of giant black holes. These black holes have an internal density of matter that is substantially less than $4 \cdot 10^{17} kg \cdot m^{-3}$, the maximum density that we have been able to measure of any object.

Pulsars

The maximum mass of a white dwarf is about 1.2 solar masses. This appears to be about the minimum mass of a pulsar. The density of a pulsar is much higher, and the radius is much smaller than that of a white dwarf. In SUiTe, a pulsar is formed when the gravitational pressure is strong enough to overcome the electrostatic pressure from protons repelling each other. The result is similar to a giant nucleus where the outer surface is constructed of shells as described in chapter 18. As the pulsar grows, the gravitational pressure forces electrons to combine with protons to form neutrons, ending shell formation.

The usual description of a pulsar is a rotating neutron star. Black holes, in the present theory of cosmology, are neutron stars that have collapsed to a singularity. In SUiTe, black holes are neutron stars, and pulsars are similar to giant nuclei. Unlike in an atom, the electrons associated with the protons in the shells are unable to move away from the surface of the pulsar because of gravitation and electrostatic attraction. Below the surface layers of the pulsar which are arranged like the shells of large nuclei, the inside may be made up of neutrons or a mixture of neutrons and protons that behaves like a superfluid due to the mutual repulsion of both the protons and the neutrons. Gravitation enables the formation of shells at and near the surface by offsetting electrostatic repulsion. The spinning of the pulsar, along with the motion of the interior nucleons, can disrupt inner shell formation resulting in free protons. These inner electrical currents resulting from the motion of the protons can affect the electron-driven external electrical current resulting from the surface electrons surrounding a rapidly spinning pulsar.

The formation of shells helps to hold the pulsar together and combines with the gravitational force to keep the radius small. The following table provides the neutron star radius that would occur in the absence of shell formation, r_{bh}, and the radius resulting from the assumption that the density is $4 \cdot 10^{17} kg \cdot m^{-3}$, r_{pulsar}. R_S is the Schwarzschild radius. Mass is in solar masses. N_c and d_{min}/d_n are necessary to determine the value of r_{bh}. The radii are in meters. Please note that $r_{bh} = N_c \sqrt[3]{4\pi \div 3} \cdot (d_{min}/d_n) \cdot 10^{-15} \cong d_{min}/d_n \cdot r_{pulsar}$.

mass	$N_c(10^{15})$	d_{min}/d_n	R_S	r_{pulsar}	r_{bh}
1.2	6,981	1.5753	3,545	11,251	17,727
1.4	7,349	1.5854	4,136	11,844	18,781
1.6	7,683	1.5951	4,727	12,383	19,756
1.8	7,991	1.6048	5,317	12,879	20,672
2.0	8,277	1.6143	5,908	13,339	21,539
2.2	8,544	1.6234	6,499	13,770	22,359

I have used a density for pulsars of $4 \cdot 10^{17} kg \cdot m^{-3}$. The black hole radius, r_{bh}, is the radius which would exist without the shell structure that exists in pulsars.

Neutron Stars

The largest pulsar at the time of this revision is 2.17 solar masses. I will assume that no pulsars will be found with 2.3 solar masses or more. I have included a column for the Einstein factor.

$mass$	$N_c(10^{15})$	d_{min}/d_n	R_S	γ_E	r_{bh}
2.3	8,671	1.6279	6,794	1.1940	22,754
3	9,474	1.6591	8,862	1.2401	25,338
5	11,233	1.7446	14,771	1.3709	31,590
10	14,152	1.9545	29,541	1.7214	44,588
20	17,831	2.4085	59,082	2.6237	69,229
50	24,201	3.8710	147,706	6.7562	151,015
100	30,491	6.0327	295,413	16.403	296,515
1000	65,692	27.926	2,954,127	398.55	2,957,223

Although the radius never shrinks below the Schwarzschild radius, for very large bodies of a million solar masses or more, the difference between the Schwarzschild radius and the radius described here becomes infinitesimal, and following Einstein, seconds become extraordinarily long. On our time scales, inside the radius, among the neutrons that are increasingly further from each other, trapped in space, nothing is happening. The situation at the surface is different. Here light is orbiting the massive body. Dark matter particles falling in are ripped apart by the light pressure at the surface and turned into their four pieces: an electron, a positron, and two neutrinos. They are ejected with considerable force, and subsequently recombine into dark matter particles producing a steady supply of photons in the halo around the black hole. Oddly enough, many of these photons have the same frequency as those expected from Hawking radiation. Black holes are excellent sources of light energy.

Chapter 6

Dark Matter and Cosmology

The hoped for possible masses for dark matter depend upon the particular branch of physics that is seeking to determine the mass. For particle physicists, dark matter must either be very small, an axion or something similar, or very large, a WIMP (Weakly Interacting Massive Particle). In both cases it is necessary to assume exotic properties for dark matter. Axions should result in a fairly homogeneous density for the cloud of dark matter in a galaxy because their small mass should correspond with a large velocity making them "hot" particles. Neutrinos are hot particles and they do not have the density profile around galaxies that dark matter has. Due to their large masses, WIMPs should rapidly condense around matter. This isn't seen either. One way to prevent this condensation is to have them slide past matter and each other as if they were a superfluid. Helium superfluids exist, but only at temperatures lower than those found in space. Neither axions nor WIMPs match the data collected by astronomers.

Dark matter was discovered by noting that galaxies had flat rotation curves—orbiting bodies around the center of the galaxy were moving at the same speed independent of their distance to the center. For this to occur, the mass contained in the sphere at radius r divided by r must be constant. The density is decreasing too quickly for axions and too little for WIMPs. A solution is to choose a "warm" particle, a particle between 10 keV and 100 keV, or a cold particle from 100 keV up to 1 GeV. It is possible in SUiTe to show that particles larger than 300 keV or smaller than 100 keV would create problems. Dark matter would continue to build up around the Sun without limit if dark matter particles were larger. If dark matter particles were smaller, the minimum temperature in the chromosphere would be lower.

In the ΛCDM model of cosmology, dark matter should be non-baryonic, cold, unable to radiate away kinetic energy, and only able to interact with normal matter through gravitation and, possibly, through the weak force. In SUiTe, dark matter is leptonic. At 162 *keV* it is a cold particle. It cannot radiate away energy, but it does interact with both light and matter by colliding with them. This makes the dark matter particle of SUiTe a very different particle than the dark matter particle of either the standard model or the present model of cosmology.

Before seeking to understand how dark matter can influence galaxies and solar systems, it was necessary to describe how it interacted with matter in the section on repulsion scattering in the previous chapter. We don't find dark matter deep in the Earth where we find neutrinos. It loses momentum quickly trying to penetrate the crust of the Earth. The heat of the Earth pushes it toward the surface even as gravity attempts to draw it down. The electron is about three times more massive than dark matter. In spite of its large mass compared to a neutrino, dark matter is even harder to measure. To understand how dark matter interacts with matter and light, we can start with an unsolved problem: coronal heating.

6.1 The Chromosphere and Coronal Heating

Unless we want to make up unusual rules for the behavior of dark matter, some dark matter will be falling into the solar system attracted by gravity. We can see no sign of its influence, so it would be reasonable to assume that the solar system is in equilibrium with dark matter. The infall of dark matter should then be equal to the ejection of dark matter. For extremely light particles, as long as they interact with light, and if the scattering cross section is large enough, light pressure from the Sun will send them back out of the solar system. The problem is that the solution will not be a universal one. The parameters that favor the Sun may not be the appropriate ones for a star that is much hotter or much colder. We might be the lucky ones. Weak stars might have a huge buildup of dark matter in their solar systems, but measurements suggest that dark matter is reduced in star clusters, and that this does not depend strongly on the star types within the clusters. This suggests that stars are not only successful at preventing a buildup of dark matter around themselves, but that they regulate the density of dark matter within their cluster.

If dark matter particles are extremely heavy, WIMPs of a *TeV* or more, they would need to pass right through the solar system because there is no source of energy that could eject them. A number of underground experiments have been searching for such particles for many decades and have come up empty-handed.

Assuming that the minimum temperature in the chromosphere corresponds with the kinetic energy necessary for a particle of dark matter to escape the Sun's gravitation, the mass of dark matter can be found using the Boltzmann constant.

$$m_{dm} = 2k_B T_{min} \div v_{esc}^2$$

where T_{min} the minimum temperature of the chromosphere. For $T_{min} = 4000$, $m_{dm} = 162\ keV$. Using this value, it is possible to determine if there are processes in the solar system that can remove dark matter at the rate at which it is falling in.

If the interstellar medium (ISM) starts at 120 AU (astronomical units) from the Sun, and if the dark matter density in the ISM is roughly 0.4 GeV/cm^3, then the infall per square meter per second can be determined by multiplying the density by a velocity. If we use the escape velocity at 120 AU as the infall velocity, then the infall is roughly $2.74 \cdot 10^{-18} kg \cdot m^{-2} s^{-1}$. In terms of dark matter particles of 162 *keV*, this would be 9,500 billion particles per second per meter squared or about $3.85 \cdot 10^{40}$ particles in total per second coming into the solar system. At the surface of the Sun, there would then be about $6.3 \cdot 10^{21}$ particles of dark matter per meter squared per second. This would result in a gain of about $3.5 \cdot 10^{17}$ kilograms per year. In one billion years, the Sun would have added less than 0.02 percent of its present mass. In terms of modifying the gravitational behavior of the Sun in a way that we could detect, there is no need for the Sun to eject dark matter, but the increase in the thickness of a blanket of dark matter around the Sun should be detectable.

In SUiTe, dark matter particles interact with light. The interaction can be treated either as Compton scattering or as elastic scattering. The maximum of 1D Elastic scattering provides about four times more energy than the maximum energy for Compton scattering. It is unlikely that the interaction of light with dark matter will ever be close to the maximum for either 1D scattering or Compton scattering. There will have to be many exchanges of energy by the same photon with multiple dark matter particles in the initial 400 kilometers of the chromosphere given the number of photons leaving the Sun.

The infall of dark matter into the solar system does not provide the densities of dark matter necessary to interact with solar photons multiple times unless it piles up in a blanket around the Sun. Due to matter-to-matter repulsion and the density of matter in the Sun, combined with heat from within the Sun, dark matter will be in a layer above the surface. Some of the energy from the Sun, in the form of light pressure, will be transferred to dark matter in this layer, consisting of the 100 km thick photosphere and the 20,000 km thick chromosphere. Many of the dark matter particles will then move outward from the temperature low-point in the chromosphere, having reached the evaporation temperature for dark matter. The evaporation temperature that will be used for dark matter is the temperature associated with the Sun's escape velocity for a particle of that mass. Some of these evaporating particles will succeed in ionizing hydrogen in the gas that surrounds the Sun and drop back down, but most will leave the solar system. The energy needed to ionize enough hydrogen to replace those particles lost in the solar wind is less than 1 watt per meter squared. The energy needed to eject as many dark matter particles as are falling in is about 335 watts per meter squared.

Although a dark matter blanket in the inner chromosphere can be made thick enough to absorb as much energy as is needed to heat the corona, the low matter density of the corona and the small scattering cross section with dark matter make dark matter an impossible source of coronal heating. But dark matter can be the source of the ions found in the outer chromosphere and the inner transition region. The rapid heating in the transition region will ionize those atoms that have not already been ionized by their interactions with dark matter.

The cross section between dark matter and photons is about $2.5 \cdot 10^{-30}\ m^2$. The maximum energy for Compton scattering with a photon from the Sun, using a wavelength of 950 nm, is only about $1.7 \cdot 10^{-24}$ joules. This wavelength comes from the assumption that the Sun radiates like a black body, a Sun surface temperature of 5778 K, photons emitted per square meter per second at the surface of the Sun of $2.9 \cdot 10^{26}$, and $60.5 \cdot 10^{6}$ watts emitted per square meter. Any given collision is likely to exchange far less energy than $1.7 \cdot 10^{-24}$ joules, requiring a single photon to strike many dark matter particles on its passage through the chromosphere. This first problem can be taken care of by having a dark matter chromosphere layer thick enough to trap the necessary energy needed to eject as much dark matter as is falling in to the solar system. An example of the possible numbers of hits between dark matter and photons that depends on the amount of energy transferred from the photon to dark matter is given below. The less energy transferred per collision, the thicker the dark matter blanket must be.

Normally, we would imagine that the chromosphere should behave like an atmosphere, with excited hydrogen and helium moving outward, cooling, and then falling back toward the Sun under the influence of gravity. Instead, the temperature of the chromosphere continues to increase from a low of about 4000 degrees after 400 kilometers, rising to 25,000 degrees at a distance of about 20,000 km from the photosphere as the chromosphere gives way to the transition region. The temperature reaches between 500 thousand and one million degrees in this region and then up to 3 million degrees in the corona. Although the most commonly cited sources of the necessary energy are pressure waves called Alfvén waves combined with magnetic field collapse and/or reconnection, the pressure waves are far too weak and the magnetic energy transfers far too inefficient to heat the transition region and the corona up to the temperatures that we see. This explanation also fails to explain the odd behavior of the chromosphere. Even perhaps more interesting is that, when it is possible to measure the thickness of the chromosphere in nearby stars, we find that its thickness in comparison to the Sun can be estimated by taking the cube root of its mass in comparison to the Sun's mass.

The Chromosphere

The need to remove as many dark matter particles from around the Sun as are falling in to the solar system explains both the existence of the chromosphere and its unusual properties. The most interesting property of the chromosphere is the rapid cooling from 5,778 K to 4,000 degrees kelvin. Why 4,000 degrees? Why not less or more? Like water on a stove, as dark matter is heated next to the photosphere, it moves away and colder dark matter comes in to replace it. Dark matter heats until it reaches 4,000 degrees at which point it evaporates. Evaporation for dark matter is interpreted here as reaching the Sun's escape velocity. 4,000 degrees would then place a limit on the mass of a dark matter particle resulting in the equation for the mass of dark matter given earlier as equal to 162 keV.

If dark matter was 0.5 MeV, about the mass of an electron, the average speed at the low temperature point of the chromosphere would be 343 kilometers per second. There would not be enough dark matter particles in the higher velocity tail of the energy distribution to prevent dark matter from continuing to build up around the Sun. If the mass of dark matter was 50 keV or less, a warm particle, the average velocity would be 1,085 kilometers per second. Dark matter would evaporate at a lower temperature than 4,000 degrees. Dark matter should then have a mass between 100 and 300 keV to match the low temperature in the chromosphere of 4,000 K. 100 keV is still heavy enough that it won't penetrate deeply into stars or planets.

Whatever mass dark matter has, the scattering cross section with normal matter will never be enough to transfer sufficient heat to the corona. The solution has to be magnetic fields. The problem is that magnetic fields are inefficient in transferring energy using neutral particles such as hydrogen molecules or helium. If there were free electrons and protons in the outer portion of the chromosphere, magnetic fields would be able to rapidly accelerate them and provide the necessary energy that we see in the transition region and in the corona.

We can imagine dark matter just beyond the photosphere as a convective fluid. After 400 km from the outer edge of the photosphere, due to contact with those particles of dark matter that have been heated to escape velocity, normal matter particles become hotter than those closer to the photosphere. There are no cold normal matter particles that can fall back toward the photosphere. Normal matter cannot cool the inner chromosphere by convection. Close to the Sun, the gas is about 74 percent hydrogen and 24 percent helium. They both radiate very well, but the radiation loss is not nearly enough to cool the inner layer of the chromosphere. There has to be something else that is able to transfer energy away from the inner chromosphere by convection. Some evaporating dark matter particles will collide with normal matter in the chromosphere beyond the first 400 kilometers, cooling themselves enough to fall back toward the photosphere. Added to those dark matter particles that cool themselves through collisions with normal matter or other dark matter particles are the infalling particles of dark matter coming from outside the solar system that reach the Sun. Dark matter particles that reach the Sun will lose their kinetic energy in a thin dense layer of normal matter just inside the photosphere, and will then be pushed back to the surface of the Sun by heat energy.

The dark matter blanket must absorb at least 335 watts per square meter to be able to eject the dark matter that is falling toward the Sun. The number of times a photon will hit a particle of dark matter depends on the scattering cross section and the particle density of dark matter in the 400 kilometers above the photosphere. The output of the Sun is about 60.5 million watts per meter squared. On average, photons must lose $5.5 \cdot 10^{-6}$ percent of their energy to provide the approximately 335 watts necessary, about $1.55 \cdot 10^{-24}$ joules per averaged photon.

Assuming a maximum energy transfer of $1.7 \cdot 10^{-24}$ joules via Compton scattering, a photon of $2.09 \cdot 10^{-19}$ joules (a wavelength of 950 nanometers) will need to strike dark matter once over a distance of 400 km. This would require a dark matter density of $1 \cdot 10^{24}$ particles per meter cubed assuming a scattering cross section of $2.5 \cdot 10^{-30}$ per meter squared. If we assume that a maximum transfer of energy is unlikely, and instead assume that a photon must strike as many as 1000 particles of dark matter in the first 400 kilometers of the chromosphere in order to lose $5.5 \cdot 10^{-6}$ percent of its energy, the dark matter density will need to be about $1 \cdot 10^{27}$ particles per meter cubed. Even doubling this value to take into account the amount of dark matter in the remaining part of the chromosphere after the temperature low point at 400 kilometers, this larger density represents only 0.7 billionths of the mass of the Sun.

Conclusion

Dark matter particles need not be treated as having properties that no other particles have. They can fall toward the Sun like any other gravitationally attracted particle. Their presence creates a blanket of dark matter around the Sun that results in the chromosphere. A small portion of the energy from the photosphere is converted to dark matter kinetic energy. Dark matter convection keeps the inner chromosphere cool. The continuing ejection of dark matter particles helps to heat normal matter in the outer chromosphere by ionizing some of the atoms. The resulting ions, in combination with powerful magnetic fields, are able to complete the ionization of all atoms in the transition region. The high temperatures in the outer edge of the transition region and in the corona are the result of magnetic fields accelerating these ions. The hottest particles in the corona form the slow solar wind. SUiTe suggests that the temperatures and sizes of the photosphere, chromosphere, and the inner transition region are all dependent on the infall of dark matter and the mass and temperature of the Sun. The temperature and size of the outer transition region and the corona vary with changes in the Sun's magnetic fields.

Both the lowest temperature in the chromosphere and its distance from the photosphere are determined by three factors which vary depending on location and the star type: the surface temperature of a star, its mass, and the number of dark matter particles arriving at the star surface per second. The mass of a dark matter particle is also important, but it does not vary. The percentage of energy from the photosphere transferred to dark matter in the chromosphere will depend on the thickness and depth of the dark matter blanket. No matter what type of star it is, the equilibrium is reached when the number of dark matter particles falling into the star equals the number ejected from the star's gravitational field.

6.2 Cusp Versus Core and Prolate Halos

The Navarro-Frenk-White model of the density of dark matter within a galaxy has been successful in modeling the motion of stars in a galaxy outside of the center, but requires that the centers not be cored. The stars in cored galaxy centers do not have flat rotation curves, but, instead, velocities of stars decrease with distance from the center up to a distance which appears to depend on the density and the brightness of the stars in the center. This unexpected result suggests that in active galaxy centers, the dark matter density is lower than expected.

When we consider the previous section on coronal heating, we can guess that most particles of dark matter that are sent out of the solar system will not stop at the edge, but will keep going, with some even leaving the galaxy. The Sun regulates dark matter in the solar system by sending dark matter out of the solar system. At a much larger level, the Milky Way is also regulating dark matter within the galaxy.

Imagine the beginning of a spiral galaxy when particles of matter and dark matter collapse into a spinning disk. At this early point, the dark matter halo of the galaxy is oblate. After stars form they eject dark matter evenly in all directions, depleting the dark matter in the original disk and increasing it elsewhere. The dark matter distribution in the galaxy moves from being oblate to being spherical. But not quite spherical, because the spherical spread of dark matter is from each star. Many of the stars are in the disk and not in the center. The stars in the disk add more dark matter to the halo around the galaxy center than the stars in the center add to the dark matter halo surrounding the disk. The distribution of dark matter becomes slightly prolate around the center while still extending outward along the original disk. Given enough time, and in the absence of galactic mergers, dark matter should be arranged in a prolate distribution above and below the center that is superimposed upon a truncated oblate distribution in line with the galactic disk, truncated because dark matter particles sent in the direction opposite to the galaxy center from the stars in the disk are more likely to leave the galaxy.

Our Sun removes as much dark matter as falls in to the Sun, but if there was more dark matter at an earlier time, the present 400 kilometers from the outer edge of the photosphere until the low temperature point in the chromosphere would have been further away. More dark matter particles per second at that earlier time would have evaporated than are evaporating now. After a considerable amount of time, less dark matter would be in the interstellar medium surrounding the Sun and the nearby stars. In an active galactic center with a high density of hot stars, this process will be repeated millions of times over. The continuing stream of dark matter leaving the galaxy center decreases the local mass density of dark matter providing a central core. But there is still a missing piece. Can too much dark matter inhibit star formation?

6.3 Too Big to Fail (TBTF) – Missing Satellites

There have been many solutions put forward for the lack of more satellite galaxies surrounding the Milky Way. The stars may have been stripped out by their attraction to the center of the Milky Way. The dark matter halos around the potential satellite galaxies may have been too thin, with the result that the gravitational potential may have been too low to form stars. Rather than addressing these potential solutions to the problems of TBTF or missing satellites, I will instead consider how dark matter could inhibit the gravitational collapse of a giant molecular gas cloud.

During the gravitational collapse of a gas cloud, gravitational potential energy is turned into the kinetic energy of the particles in the cloud. For gases, temperature is a measure of the average kinetic energy of the molecules of the gas. This kinetic energy also exerts a pressure within the cloud. If the pressure exceeds the gravitational pressure driving the collapse, the cloud stops collapsing.

The loss of kinetic energy as radiation by an ideal gas is proportional to the fourth power of the temperature. This loss guarantees that the gravitational collapse will continue even as the temperature of the gas increases. When the temperature is high enough that the gas ionizes, the loss of energy through radiation greatly decreases. This is the point where collapse slows. Cooling proceeds partly through the reduced radiation loss, but also by the loss of the most energetic protons and electrons until the gas is sufficiently concentrated that the increase in the kinetic energy of the ions from the collapse is no longer sufficient to offset the force of gravity. The collapse continues until the highly concentrated ionized gas undergoes fusion.

When dark matter is added to the story, it is similar to adding ions. Dark matter can only lose energy through collisions with slower moving particles. Although the gas atoms and molecules are gaining energy via collisions with dark matter, they radiate away far more energy than they gain. If the percentage of dark matter is low enough, the collapse will continue and the most energetic particles of dark matter will escape. As long as the ejection of dark matter exceeds the infall, stars will eventually form. The greater the percentage of dark matter in the cloud, the slower the collapse becomes.

But what if the loss of dark matter fails to exceed the amount of infalling dark matter? The cloud can reach an equilibrium where the internal kinetic pressure equals the gravitational pressure. The temperature stops increasing and the percentage of dark matter in the cloud may even slowly increase. If a star can form in the cloud, it will reduce the amount of dark matter around it in the way described in section 6.1. This reduction will allow gravitational collapse to continue in that particular part of the cloud. Stars create stars. Star "nurseries" result from the first stars ejecting dark matter from the ISM surrounding them.

This is a game where the gas clouds have the best chance at forming stars during the beginning of galaxy formation because dark matter is being ejected faster than it is being added. The less dark matter there is in a part of the cloud that is collapsing, the more rapidly it can collapse. Star formation proceeds from the center of a galaxy outwards. But once the stars in the center of the galaxy and the near disk have formed, molecular clouds further away from the center will be continually bombarded by dark matter particles coming from the center. Many of these dark matter particles will be gravitationally attracted to these clouds and fall in. Like the problem with dark matter in the solar system, once the temperature of the cloud is high enough, dark matter will reach escape velocity and leave the cloud. The average temperature of these gas clouds, between 10 and 20 degrees, when applied to dark matter particles, results in a kinetic energy well below that which is necessary to overcome the escape velocity of the cloud combined with the velocity necessary to leave the external gravitational field from the galaxy center.

Unlike coronal heating, there is no buildup of a blanket of dark matter in or around the cloud because there is no central source of mass. Although normal matter in the cloud can continue to radiate away its kinetic energy, dark matter can represent more than 90 percent of the total mass of the cloud. Because the particle density is low, and the particle density of normal matter is especially low, it is difficult for dark matter to lose its kinetic energy via collisions with normal matter. This results in a lower temperature for normal matter than for dark matter. It is possible for an equilibrium to develop between the gravitational pressure and the dark matter kinetic pressure that halts collapse.

A gas cloud that is trapped in an equilibrium at a low temperature because of dark matter can nevertheless form stars if there is a sudden increase of energy in the form of light and fast moving particles of matter. It is in this way that a supernova can suddenly heat a portion of a gas cloud to the temperature necessary for dark matter particles to escape. With the shift away from the equilibrium between gravitational pressure and dark matter kinetic energy, collapse can once again begin. As soon as the temperature exceeds that needed for dark matter to reach escape velocity, dark matter can no longer prevent collapse.

If this story is true, we should be able to find gas clouds in the Milky Way that would otherwise be capable of collapsing into stars, including very large clouds that should have formed satellite galaxies, with temperatures remaining too low for dark matter to reach escape velocity. If it is possible to measure the dark matter percentage within the cloud, it should be above the average found in the ISM at that radius from the galaxy center.

Can a Cloud Made Only of Dark Matter Collapse?

If we imagine that bodies such as stars and exoplanets are gravitationally part of the Milky Way up to 210 kilo-parsecs (kpc) from the center, and if the mass of the Milky Way at this radius is 1.5 trillion solar masses of which 90 percent is dark matter, then the average dark matter density will be about $2.355 \cdot 10^{-24}$ kilograms per meter cubed. We can estimate the Jeans mass, the mass at which gravitation will cause a cloud of massive particles to collapse, by first estimating the Jean's wavelength and assuming that the wavelength is twice the radius of the cloud. In the situation here, the mass is not interesting, but the wavelength is. The wavelength is the diameter of the cloud. If the wavelength can easily fit inside the Milky Way, collapses should occur. There would be no reason for a lack of satellite galaxies, and molecular clouds would not be too big to fail.

If dark matter has a mass of 162 keV and dark matter represents 85 percent of the mass of the universe, then there will be about 33,000 particles of dark matter for every nucleon. In this case, the initial collapse of a molecular cloud will depend almost entirely on dark matter. I will consider the case of a cloud of only dark matter. A formula for the Jeans wavelength in terms of the temperature and the density of a gas cloud is

$$\lambda_J = c_s \cdot \left(\frac{9}{4\pi G \rho_{dm}}\right)^{1/2} = \left(\frac{5 k_B T}{2\pi(\gamma - 1) G \rho_{dm} m_{dm}}\right)^{1/2}$$

If we treat dark matter particles as a monatomic gas, γ is equal to $5/3$.

For a temperature of 10 kelvin, with dark matter at a mass of 162 *keV*, the wavelength is 62 kpc. Divided in two, the radius of 31 kpc is larger than the radius of the visible Milky Way, the part of the Milky Way where the vast majority of the stars are. The cloud size is not too large for an initial collapse into a galaxy the size of the Milky Way, but it is too large for subsequent satellite galaxies within the Milky Way. This suggests that a total galactic mass composed of 85 percent dark matter can form large star formations such as the central portion and disk of the Milky Way, but makes the formation of satellite galaxies within the larger galaxy difficult without some way of increasing the percentage of normal matter within a smaller region of the collapsing cloud.

It is important to note that even at 10 degrees, dark matter can easily escape a giant molecular cloud if that cloud is not part of a galaxy. The difficulty is when the cloud is subject to an external gravitational field. Even at 210 kpc from the center of the Milky Way, the escape velocity is 250 km per second, well above the average velocity of a dark matter particle in a cloud. Globular clusters and satellite galaxies in the Milky Way are more likely to have been captured than to have formed within the galaxy.

Decreasing the mass of a dark matter particle to 50 *keV*, there is a Jeans wavelength of 111 kpc, much larger than the diameter of the visible Milky Way. Increasing the mass of a dark matter particle to 1 *MeV*, we have a Jeans wavelength of 25 kpc, roughly the diameter of the visible Milky Way. At 100 *MeV*, the diameter is only 2.5 kpc. Satellite galaxies could easily form.

The range that we find here for dark matter is less constraining than for coronal heating, but it still excludes axions with masses less than 10 *keV*, and WIMPs with masses greater than 1 *GeV*. They would be impossible.

6.4 CMB and AME

The cosmic microwave background radiation (CMB) has a slightly higher temperature when measured from the southern hemisphere than from the northern hemisphere. This is the CMB dipole anisotropy. The CMB should have the same temperature in all directions, but there is a simple kinematic answer for the temperature difference; the local group of galaxies, including the Milky Way and Andromeda, are moving in the direction of a group of galaxy superclusters of which the Virgo supercluster is the closest. Light from that direction is blueshifted, hotter.

Dark matter can be used solve problems, as it does with coronal heating, but it can also create problems. Dark matter is denser near the center of the Milky Way. We know this simply by looking at the circular velocity of the stars orbiting the center. All sources of normal matter taken together are substantially less than the mass necessary for their velocity. Dark matter must fill in the rest. Overall, dark matter is estimated to be 90 percent of the mass of the Milky Way. Although the percentage of the mass corresponding to dark matter is much less when measured very close to the center of the galaxy, because the concentration of normal matter near the center is much higher than the average, we can use the value of 90 percent to help understand the problem.

At a distance of 3 kpc from the center of the galaxy, if the circular velocity of the stars there is 250 km per second, and if 90 percent of the interior mass is dark matter, and if the mass of a particle of dark matter is 162 *keV*, and if the scattering cross section of dark matter with a photon of any wavelength is about $2.5 \cdot 10^{-30}$ square meters, then the number of times a CMB photon will be scattered by a particle of dark matter over a distance of 1 kpc is 6.3. At the Sun's distance from the center of the galaxy, a CMB photon will be scattered 0.74 times by a particle of dark matter over a distance of 1 kpc.

The formula for the dark matter particle density at a distance of *d* from the galaxy center, where the circular velocity is *v*, dark matter represents 90 percent of the mass at a distance *d*, and the volume of space at the distance *d* is V_d, is:

$$n_d = \left(0.9\, {}^{M_d}\!/_{m_{dm}} \div V_d\right) = 0.9\left(v \div \sqrt{{}^{G}\!/_{d}}\right)^2 \div (m_{dm} \cdot V_d)$$

The approximate probability that a CMB photon will be scattered by a dark matter particle traveling a distance of 1 kpc where the average dark matter particle density is as defined in the above formula in the neighborhood of the distance *d* from the center of the galaxy is:

$$P_S \approx 100 \cdot n_d \cdot \sigma \cdot kpc$$

The remaining question is how much energy will be transferred in a collision between a dark matter particle and a CMB photon, if the dark matter particle is orbiting the center of the galaxy at the same velocity as the stars. This would be similar to inverse Compton scattering, though the methods used for inverse Compton scattering would not work for dark matter. If we use 1D elastic scattering to provide a ballpark estimate for the maximum energy transfer, the transfer that would occur if a CMB photon was reflected by a dark matter particle, would result in a change in temperature of the CMB photon of 0.009 degrees using a CMB average energy of 2.725 K.

That's where the problem comes in. Even though the maximum energy transfer may well be 1000 times the average energy exchanged between a dark matter particle and a CMB photon, along a line from the Sun that is tangent to a 2 kpc circle around the galaxy center, a CMB photon will collide more than 100 times with particles of dark matter. There should be some sign of this.

The CMB dipole does not point at the galaxy center, but if the quadrupole and the octupole both point in the direction of the galaxy center, this secondary effect is likely due to the dark matter-CMB interaction. If not, there is a difficult-to-answer question as to why there is no effect on CMB photons that pass near the galaxy center on their way to our telescopes. If dark matter gains energy from photons in the chromosphere, it must also interact with all photons, including CMB photons.

AME

The careful study at multiple frequencies of radiation emitted by gas clouds has found an anomalous microwave emission (AME) between 10 and 60 gigahertz. The most probable cause is spinning particles of dust where the particle of dust is an electric dipole. If a particle of dust absorbs radiation, it will vibrate.

Some of the vibrational energy will be turned into angular momentum. Because the spinning particle is an electric dipole, it should radiate away energy according to Larmor's formula. The energy radiated away as AME will cause the dust particle to stop spinning if no additional energy is added through collisions or radiation absorption. Spinning dust particles are commonly found in condensing molecular clouds and in star-forming or planet-forming regions.

The difficulty is finding a source of energy sufficiently powerful and continuous to cause the dust particle to spin-up and to remain spinning. If the radiation source is the cloud itself, the temperatures involved would be ionizing. A better source of energy would be matter-to-matter collisions. Here the problem is an insufficient density of gas molecules and dust particles to maintain the AME even if some dust particles have already spun-up. Dark matter could potentially resolve the problem of the energy source.

If dark matter is responsible for TBTF and missing satellites (previous section), then it will only be able to lose the energy it has gained through gravitational collapse by colliding with normal matter particles in the molecular and dust cloud. With TBTF and missing satellites, the collapse of the cloud is stopped early at a low temperature by dark matter pressure. Loss of dark matter energy through collisions is ineffective due to the low density of the cloud. In the case of AME, the collapse is proceeding and the density is high enough that dark matter can more easily transfer its kinetic energy to normal matter. The source of the energy is inexhaustible, because, as energy is radiated away by normal matter, the cloud continues to collapse, adding energy to the dark matter particles, and replacing the kinetic energy that dark matter had lost through collisions.

Chapter 7

Electrostatic Force

The gravitational force is not due to attraction, nor is magnetism due to attraction or repulsion. Both of these forces are more complicated. The electrostatic force is one of the three fundamental forces and has both attraction and repulsion. Because the electrostatic force is completely canceled in the particles that make up space, the electrostatic force is not affected by space. It is not possible to create an electrostatic wave in space. This contradicts the usual interpretation of electromagnetism where an electromagnetic wave in space consists of both electrostatic and magnetic components that are in phase and at right angles to each other.

This is not the only disagreement with classical electromagnetism. Even particles as small as an electron are believed to emanate the electrostatic force in all directions equally, much as gravitational sources were believed to do prior to Einstein's theory of general relativity. Instead, electrons emit the electrostatic force along a line. The direction of this line is important in moving from classical electromagnetism to quantum electromagnetism.

In both the standard model and in SUiTe, the electrostatic force decreases with the inverse-square of the distance, the usual inverse-square law that applies to gravitation as well. Here again, the reason is different. The force of the electrostatic forceline of the electron does not decrease with distance, but the forceline interacts with more and more other electrostatic forcelines the farther from the electron the electrostatic forceline is measured. These additional external forcelines weaken the forceline of the electron through screening.

The model that has been used, but which is incorrect, is that of the Sun from which photons are flowing out in all directions roughly equally. If we surround the Sun with concentric spheres, we will note that even though the total number of photons remains the same measured at the surface of each sphere, the number of photons per square meter follows the inverse-square law. This model, which is accurate for photons streaming from a light source, is inaccurate for the electrostatic force.

With screening from the background electrostatic field, which is made up of electrostatic forcelines from all charged particles throughout the universe, we once again use concentric spheres around the charge source we wish to measure. Screening is dependent on the number of forcelines that penetrate the sphere from the outside. Background electrostatic forcelines that are opposite in charge mask each other. Electrostatic forcelines of the same charge repel each other. Two forcelines with the same charge cannot enter the sphere at the same location. The number of forcelines from outside the sphere that can screen the charge inside the sphere depends on the area of the surface of the sphere. In this way, increasing the distance from the charge source increases the number of screening forcelines interacting with the charge source according to the inverse-square law.

The importance of this re-interpretation can best be understood by considering the helium atom. If I combine two electrons with an alpha particle, the nucleus of helium, I will have a helium atom. I will also have turned off almost all the electrostatic charge of the two protons and the two electrons that make up helium. I cannot install two light detectors at opposite sides of the Sun and turn off the light from the Sun, nor can I add another Earth opposite ours and turn off the gravitation of the Sun. The electrostatic charge does not emanate out from a charge source in all directions, but acts along forcelines.

7.1 The Electron

In the appendix, I describe the construction of the electron. It was originally made—prior to the beginning of our universe—from three separate particles that fused together. Each of the three particles had an electrostatic forceline and a weak-force forceline at right angles to each other. Two of the three particles that make up an electron are negative, and one is positive for both the electrostatic force and the weak force. In the case of the electrostatic force, the binding of the one positive electrostatic forceline with the two negative electrostatic forcelines left what was essentially a single negative electrostatic forceline. The three weak-force forcelines are not as close to each other in the electron and interact separately with the particles around them. This interaction results in the property called "spin." Spin is discussed in more detail in chapter 14 on the Stern-Gerlach apparatus and in chapter 16 on the two-slit experiment and Bell's inequalities.

The electron attempts to move in an electrostatic field by aligning its combined negative electrostatic forceline with the field. You can imagine something in the shape of a cylinder with rounded ends aligned lengthwise in the field and moving in the direction that the electrostatic forceline is pointing. The electrostatic forceline is parallel with the longer axis of the cylinder.

Like all particles, including the particles of space, there is nothing about an electron that could be called solid. Each electron is a bundle of forces. We assign a specific location for the electron at the locus of the bundle, but that is simply a convenient choice. The influence that the particle's forcelines have on surrounding matter and space extends well beyond the position assigned to the electron. The electrostatic forcelines of the electron contribute to the electrostatic field of the universe. The electron acts on and is acted on by all other electrostatically charged particles.

Though electrons cannot be considered as having a location at a point, it is wrong to imagine that electrons cannot be assigned a location. Electrons, excluding the extensions of their electrostatic and weak-force forcelines, can be enclosed in a sphere of about 1.2 femtometers in diameter. They are not smeared out along the surface of a sphere surrounding a proton in a hydrogen atom. The electron can be anywhere on the surface of a sphere surrounding a proton in a hydrogen atom, but the electron is always in one place at a given time. Our inability to find that location without disturbing the electron does not mean that the electron is in multiple locations simultaneously until measured.

The larger the number of electrons in an atom, the more restricted the location of an individual electron is. When molecules form through the exchange of a valence electron, the valence electron is always between the two atoms. Electrons are in constant motion above absolute zero, but this motion involves random movements around a central point, random movements that increase in intensity and in distance from that central point with increases in temperature.

Electric charge is conserved. It is an intrinsic property of every particle, whether charged or neutral. Every neutral particle is constructed from even numbers of oppositely charged particles. Because charge is conserved and because our universe is three-dimensional, there is a law, originally due to Gauss, for electricity. The units used in the following equation are Gaussian.

$$\nabla \cdot \boldsymbol{E} = 4\pi\boldsymbol{\rho}$$

Here $\boldsymbol{\rho}$ is the charge density, the number of charges in a volume of space. If we imagine enclosing the source of electric charges within a sphere that has a radius equal to 1, then $4\pi r^2 = 4\pi$ is the surface of that sphere. Gauss's law states that all the charges that are in a volume enclosed in a sphere will be visible outside of the sphere at whatever distance we choose for the radius of the sphere. This equation also states that the total charge felt by a test particle outside of any sphere surrounding the charges will be subject to the inverse-square law.

7.2 Coulomb's Law

The force between two charged objects depends directly on the total charge of each object, and inversely on the square of the distance between them. Mathematically we have:

$$F = k_C \, {q_1 q_2}/{r^2}$$

where k_C is the Coulomb force constant, q is the charge, and r is the distance between the two charged objects q_1 and q_2.

This is the first clear example of quantization in physics. Charged objects carry whole numbers of charges. There is no such thing as a partial charge. q_1 and q_2 represent integers multiplied by the electric charge. The signs of q_1 and q_2 are opposite if the force is attraction, and the same if the force is repulsion.

The value of the Coulomb force constant cannot be derived from theory, but is dependent on the inverse ratio of free electrostatic forcelines to free weak-force forcelines. At roughly the distance between two of the particles of space that exhibit the weak force, about 10^{-18} meters, the weak force and the electrostatic force reach parity. Even within the nucleus, in the range of one femtometer, the electrostatic force is measurably weakened in comparison to the weak force. The electrostatic force constant is not constant at distances where the weak force is no longer effectively screened out by space. At sufficiently small distances, at 10^{-18} meters, the electrostatic force is no longer strong and the weak force is no longer weak.

Chapter 8

Magnetic Force

The magnetic force is a weak-force phenomenon involving the particles of space that have two weak-force negative forcelines, the w-duals. A spinning electron causes w-duals to line up end-to-end and spin in the same direction with each other. This change to the w-duals surrounding a magnet allows the spinning electrons in the magnet to influence other objects nearby. The arrangement and spin direction of the w-duals constitute the magnetic field. The strength of the magnetic field is determined by the density of the threads of spinning w-duals. All of the threads are anchored to spinning electrons.

There are two duals that make up space—the w-dual and the d-dual—but the d-dual is only necessary in determining the speed of light, in determining the rotational speed of the w-duals, and in playing a role in gravitation. For electromagnetic phenomena, we can ignore the d-dual.

The w-duals normally point toward matter, to which they are attracted, and spin either clockwise or counterclockwise with no preference for either direction. All particles of space have no mass in and of themselves but can exert pressure on particles that do have mass. The attraction of w-duals to matter is an important part of gravitation. The gravitational force is caused by the decrease in the pressure of space between two objects as compared to the pressure of space elsewhere surrounding the two objects.

When w-duals are compressed between two charged particles, they exert pressure on both. This property is responsible for keeping electrons out of the nucleus in atoms. A photon is a moving compression of the w-duals traveling at the speed of light similar to the way a phonon, a moving sound compression in a crystal, travels at the speed of sound in the crystal and is able to transfer sound energy. We describe the compression of this pulse made up of w-duals in terms of its wavelength. The shorter the wavelength, the more powerful the pulse, and the more powerful the pulse, the more pressure it can exert on matter. A fourth way that w-duals exert a force on matter is through magnetism.

The structure of an electron can be thought of, for the purposes of magnetism, as being similar to a plus sign. One axis carries the electric charge. When the electron spins, it spins around this axis. The other axis is the negative of the weak force. As the electron spins, the w-duals next to the electron line up parallel to the electrostatic forcelines of the electron and spin in the opposite direction that the electron is spinning. This circle of w-duals surrounding the spinning electron creates a thread of w-duals which extends both in front of and behind the electron. Magnetic fields are made up of such threads. Many of the threads are anchored by a single electron, but it is also possible that a thread can have more than one electron involved. In that case the thread will be able to exert some integer times the force of a thread that corresponds to a single spinning electron.

One magnet can magnetize an object made of a magnetic metal such as iron placed in its magnetic field, because some of the threads line up with electrons that are free to begin spinning in the non-magnetized metal. Once the free electrons in iron begin spinning, the iron becomes temporarily magnetized. If the external field is strong enough, and it lasts long enough, iron can become permanently magnetized. This requires a movement of the atoms in the iron, so that individual electrons in a series of iron atoms are able to line up with each other. The necessary transfer of energy required to move and spin-up the electrons in the previously non-magnetized iron reduces the strength of the external magnet.

Magnetism is not an intrinsic property such as electric charge. I can hit a piece of iron that has been slightly magnetized with a hammer and some or all the magnetism of that piece of iron will disappear. If I heat an object that has been magnetized to a sufficiently high temperature, all the magnetism will disappear. The heat causes the spinning electrons to move erratically, and they cease to all spin in the same direction which is necessary for a magnet. Natural magnets—lodestones—are the result of magnetic materials cooling in the presence of the Earth's magnetic field. As the natural magnet cools, the atoms rearrange and some of their electrons align and spin with the external magnetic field. The electric charge of an object is not affected by anything we can do to the object. It is intrinsic. Magnetism is a mechanical property.

As explained in the last chapter, electric charge follows the inverse-square law. This follows from the conservation of charge and a three-dimensional universe. The source of the electric field is the charged particles. There is no permanent source for a magnetic field. A spinning electron is a temporary source of a magnetic thread in a magnetic field, but the moment it stops spinning, the thread is gone. If all the electrons stop spinning the magnetic field disappears.

The other important feature of a magnetic field is that the threads spread outward both in front of and behind the spinning electron. The arrangement of the w-duals surrounding matter is like pins sticking out of a pincushion. Even though the w-duals lack mass and can be made to flip their spin from clockwise to counterclockwise without radiation loss, the imposition of a magnetic field requires the w-duals to move. Magnetized space has a different structure than non-magnetized space. We can see the signs of the tiny threads that weave their way through space away from and back to a magnet when we sprinkle iron filings around a bar magnet and see similar patterns that are made up of huge numbers of individual threads.

Gauss's law for magnetism is similar to his law for the electric field, except that the electric field continues to spread through space no matter how far the field is from the source. The threads of a magnetic field loop back to the spinning electrons that are the source of the field. To explain this mathematically, we can say that the divergence of the magnetic field (*H*) is zero in all directions:

$$\nabla \cdot \boldsymbol{H} = 0$$

The threads that leave the source that were created by the spinning electrons in the source return to the same spinning electrons to attach to the opposite side of the electron. Magnetic threads are loops.

8.1 Properties of Magnets

If an electron is spinning clockwise when we look at it from behind, then it is spinning counterclockwise when we look at the same electron from in front. And if an electron is spinning clockwise when we look at it from behind, then the direction we are looking at is toward the north pole of the magnet. If the electron had been spinning counterclockwise, then we would be looking in the direction of the south pole. Although the decision to name one direction of the magnetic field emanating from a magnet "north" and the opposite side "south" was arbitrary, the relationship between the poles and the direction of the spin of the electron is not. I will be able to show in the next chapter that once the choice of names had been made, there was no choice in the direction of spin associated with the name.

The minimum unit of a magnetic field is a spinning electron. It is the quantum of the magnetic field. The speed with which an electron can spin is determined by the speed at which w-duals spin, which is, in turn, determined by the relative density of w-duals to d-duals—the same ratio that determines the speed of light. Because each spinning electron has a north and a south pole, depending on whether we are looking from behind or in front, it is impossible to divide a magnet such that it has one pole but not the other.

Looking again from behind the electron toward the north pole with the electron spinning clockwise, the weak-force forcelines of the electron, like a propeller with two blades, are causing the w-duals surrounding it to spin counterclockwise. Imagine now that there is a second magnet in front of the first magnet with the electrons spinning clockwise in the first magnet. Imagine that the electrons in the second magnet are spinning clockwise as well. The w-duals of both magnets will be spinning in the same direction and will not interfere with each other. This is the case when two magnets attract. In both magnets, the north pole was in front of the spinning electrons. We know that behind the electron in the second magnet is the south pole. From this we can conclude that magnets attract each other if the north pole of one magnet is placed in line with the south pole of another magnet.

If the electrons in the second magnet had been spinning counterclockwise as we looked at them from behind, then we would have been looking in the direction of the south pole of the second magnet. The north poles of the two magnets would then be closest to each other. The w-duals spinning counterclockwise in the first magnet will encounter w-duals spinning clockwise in the second. The two magnets will repel each other.

Magnetic attraction and repulsion are the result of small changes to the pressure of space between one magnet and either another magnet or, in the case of attraction, a material which can support a magnetic field. Normally, the spins of the w-duals are randomly arranged in the space between two objects. There are just as many matched pairs as unmatched pairs of w-duals. In attraction, there are more matched pairs—w-duals lining up end-to-end and spinning in the same direction. This reduces the pressure between the two objects, and they move toward each other. In repulsion there are more unmatched pairs than there would be in the absence of a magnetic field. This causes the objects to be pushed apart.

The mechanism is different than gravitational attraction, where the increased density of space between two objects causes the forcelines of the two objects to tilt toward each other, and this tilting leads to a lowering of the pressure of space between the two objects. In magnetic attraction, there is no initial tilting of the forcelines of the objects and there is no change in the density of space. Instead, there is a change in the degree to which the w-duals repel each other.

8.2 Quantum Magnetic Fields

A continuous magnetic field is a useful approximation unless we are attempting to measure the behavior of small charged particles such as electrons. To understand the behavior of an electron in a magnetic field, we have to understand that the magnetic field is constructed of threads of w-duals spinning end-to-end, and that these threads may be less than a femtometer in diameter. An electron does not enter into a continuous environment like a diver plunging into water. The electron enters into a forest of magnetic threads, with each thread anchored to a spinning electron. Moving through this forest, an electron is continually going into and out of contact with the spinning electrons that anchor the threads in the magnet. In the next chapter, I will describe electromagnetism from the point of view of an electron.

Another phenomenon of magnets is the ability of a magnet to magnetize iron when left in close proximity to iron for an extended length of time. This is not something that can be adequately explained by the standard model. A magnet is supposed to result from unpaired electrons aligning their internal magnetic fields with each other in the presence of a magnetic field. When the source of the field is removed, why wouldn't the unpaired electrons revert to their previous random arrangement? And why does time play a role in the magnetizing of a piece of non-magnetized iron?

In SUiTe, for a magnet to attract another object, it is only necessary that the other object allow the threads to pass through it. This is enough to allow space to change between the two objects reducing the pressure between them, and for the pressure of space from the outer ends of the two objects to push them together. The electrons in the non-magnetized object need not be spinning for attraction to occur. For an electron to permanently spin-up in the non-magnetized object, there must be enough energy for some of the atoms to realign and their electrons to begin spinning. If a spinning electron in the magnetized object happens to be directly in line with a non-spinning electron in the non-magnetized object, there will be repulsion—just as the attempt to match a spinning gear with a non-spinning gear results in a torque that should spin up the non-spinning gear while slowing down the spinning one. Nature does not allow different spin speeds for the electrons because of the interaction with the w-duals. It is all or nothing.

In chapter 10, on photons, I will discuss light and the emission of light by electrons moving in atoms. The emission process is an exponential decay, the same as with radioactive decay. Probability is involved. Eventually, an excited atom will decay by emitting a photon. Exponential decay is a common property of nature, especially in the world of the very small. In SUiTe, electrons are constantly being jostled by background radiation at all wavelengths. This energy can cause the electron to rotate. If enough energy strikes the electron in such a way as to cause a rotation in the direction that matches the rotation of the w-duals surrounding it, an electron in non-magnetized iron can spin up.

If electrons in iron can spin up, they can also spin down provided sufficient torque is applied and the torque is opposed to the spin of the electron. Two magnets forced together, north to north or south to south, should find their magnetic strength decreasing over time.

Chapter 9

Electromagnetism

Electrons moving in a conductor create a magnetic field. A conductor exposed to a changing magnetic field will produce a current. The ampere is the measure of current. It is the number of electrons measured per unit time passing through the conductor. If we consider the electrons passing through a wire of cross-sectional area A, the current density will be: $j = {}^{i}/_{A}$ where i is the current. We can also define current as being similar to a continuous flow of charge with a charge density equal to ρ. In that case the current density will be a vector: $\boldsymbol{j} = \rho\boldsymbol{v}$.

Analogies can break down, but if we can adjust our idea of what a velocity would mean for electrons in a conductor, perhaps the above equations create a good image. Electrons appear to boil off of a wire toward a positively charged screen in a cathode ray tube. Once the electrons leave the wire, the sense of velocity in the equation for a current density is correct. We don't know exactly what is going on in the wire. We don't know how electrons can free themselves from one atom only to be trapped by the next and still have a velocity when they leave the last atom of the conductor and move into free space. Perhaps we could consider the wire as a pressurized environment. The velocity we see at the end may be similar to a compressed gas escaping through a small hole, which is a different analogy.

Ampère suggested that the relation between a magnetic field $\boldsymbol{B}$ and a current $\boldsymbol{j}$ flowing through a wire is: $\nabla \times \boldsymbol{B} = {}^{4\pi}/_{c}\,(\boldsymbol{j})$. This equation tells us that a current in a wire creates a magnetic field around it. Why does this look familiar? If we set $k = {}^{v}/_{c}$ where we have chosen the direction of the velocity to be in the direction of the current, then we have:

$$\nabla \times \boldsymbol{B} = {}^{4\pi}/_{c}\,(\boldsymbol{j}) = k4\pi\rho = k(\nabla \cdot \boldsymbol{E})$$

The curl of the magnetic field is proportional to the divergence of the electric field generated from the charge density in a wire. In other words, the magnetic field generated by a current running through a wire depends on only two things: the number of electrons, and their speed proportional to that of light.

Maxwell made an important addition to this relation noting that electrons create a displacement electric field that must be added to the current when the electric field is changing over time.

With the correction we have: $\nabla \times \boldsymbol{B} - \frac{1}{c} {}^{\partial \boldsymbol{E}}/_{\partial t} = {}^{4\pi}/_{c}\,(\boldsymbol{j})$.

This addition by Maxwell makes the above equation anti-symmetrical with Faraday's law of induction when the current density in the above equation is 0:

$$\nabla \times \boldsymbol{E} + \frac{1}{c} {}^{\partial \boldsymbol{B}}/_{\partial t} = 0$$

$$\nabla \times \boldsymbol{B} - \frac{1}{c} {}^{\partial \boldsymbol{E}}/_{\partial t} = 0$$

9.1 Ampère's Law

How can we explain Ampère's law: $\nabla \times \boldsymbol{B} = {}^{4\pi}/_{c}\,(\boldsymbol{j})$, according to which a current through a wire creates a magnetic field around a wire? Electrons were not known to exist when electric currents were first studied. The decision was made to consider the current as the motion of a positively charged fluid toward a negative terminal connected to a source of electricity such as a battery. To describe the way in which the magnetic field enclosed the current, the thumb on the right hand represented the direction of the current, and the fingers, enclosed around the wire, indicated the direction of the magnetic field threads. It will be better for our purposes to use the left hand instead, with the thumb pointing in the direction of the motion of the electrons. The electrons are moving toward the positive terminal of a battery connected into a circuit.

A conductor such as a copper wire presents an electric field to the valence electrons of the copper atoms, where the attractive, positive side of the field is always just ahead of them in the direction of the electron flow. This is because an electron leaving a copper atom creates a positively charged "hole" that is attractive. When the electrons are moving between holes, their electrostatic-negative forceline is aligned with the current. The movement of the electron causes the weak-force forcelines that are perpendicular to the electrostatic forceline in the electron to interact with the attractive electrostatic-positive forcelines ahead of them. The interaction between two unlike forces in SUiTe results in either a push or a pull away from the plane of their interaction. This causes the electron to spin. As we determined in the previous chapter on magnetism, spinning electrons create magnetic threads by coordinating the spin and direction of the w-duals surrounding the electron.

Let's consider the electrons moving along the surface of the wire. Before the current flows, w-duals are pointing toward the center of the wire due to their attraction to matter. The weak-force forcelines sticking out from the spinning electrons moving at the surface of the wire will initially strike the w-duals end-to-end which will generate the maximum repulsion for the w-duals. Those w-duals which are hit will arrange themselves with their long axis along the surface of a ring surrounding the wire. At this point, the spins of the w-duals are still randomly aligned. The pairs of weak-force forcelines of the w-duals are perpendicular to the direction of the current. The movement of the electrons in the direction of the positive terminal causes the weak-force forcelines of the spinning electrons to brush against the two weak-force forcelines of the w-dual. The w-duals that formed bracelets around the wire as described before will be aligned end-to-end and spinning in the same direction.

If you take your left hand and point your thumb in the direction that the electrons are flowing in the wire and curl your fingers, the direction of the north pole of the magnetic field will be in the direction along your fingers toward your fingertips. In the case of permanent magnets, the direction of the spin of the electrons determined the direction of the magnetic field. In this case, the direction of the spin of the electron is unimportant. Only the direction in which the electrons are moving is important.

Along the surface of the wire, the resulting magnetic field points in one direction curling around the wire and does not interfere with the electrons creating the field. Inside the wire the magnetic threads created by the spinning electrons have no preferred direction. From this we can draw two conclusions: the net internal magnetic field is zero, and resistance is caused by spinning electrons creating magnetic fields inside the wire that interfere with each other. If we could stop the electrons from spinning as they move through the wire, there would be no internal magnetic fields and no resistance.

In chapter 13 on the Bohr atom, I will explain how the subshells form in the presence of a magnetic field. Adjacent electrons arrange their weak-force forcelines perpendicular to each other. This arrangement prevents either electron from spinning. Iron has four free electrons that have no adjacent electrons with which they can pair. An external magnetic field can then cause some of these electrons to spin up.

For two free electrons to resist spinning when moving through a wire, they will have to keep their weak-force forcelines perpendicular to each other. Such pairs are called "Cooper pairs." The less energy (heat) the paired electrons are exposed to, the easier it will be for them to remain in a paired state. From this we can predict that when we drop the temperature of a conductor below some level appropriate for that conductor, the moving electrons will be able to form Cooper pairs, and the conductor will become a superconductor. Because none of the electrons involved in Cooper pairs can spin, there is no magnetic field despite there being a current. This is a type 1 superconductor.

We also found that the magnetic field is caused by electrons along the surface of the conductor but not by those inside. If the surface of the conductor is too hot to form pairs, but the interior of the conductor can form pairs, then it will be possible to have a superconductor where a thin outside layer of electrons creates a magnetic field. This is a type 2 superconductor.

Counting the number of electrons passing through a point on a conductor would be an accurate way to measure an ampere, but it wouldn't be particularly easy. Another way is to measure the attractive force between two parallel conductors placed 1 meter apart. The formula for the force is: $F = {\mu_0 I_1 I_2 L}/{2\pi r}$. In the vacuum of space, μ_0—the magnetic permeability—is not a directly measured constant but depends on the speed of light and the permittivity of space, ϵ_0. If r, the distance between the wires, is 1 meter, and L, the length of the wire section that we are measuring, is 1 meter, and if the current supplied to each conductor is 1 ampere, then the force must be ${\mu_0}/{2\pi}$. We can now ask why two parallel conductors with currents moving in the same direction attract.

Using the left-hand rule where the thumb points in the direction that the electrons are moving, the curled fingers indicate the magnetic field curling around the wire. We can take our right hand with our thumb pointing in the opposite direction, the direction of the current, and our fingers will again curl in the direction of the field. As we bring our two hands together, we see that the w-duals indicated by the fingers of our left hand are spinning in the opposite direction from the w-duals indicated by the fingers of our right hand where the tips of the fingers of our two hands meet. When magnets attract end-to-end, the w-duals must be spinning in the same direction. The problem of w-duals side-by-side is that they are like toothed gears. If we attempt to match the spin of two toothed gears, both spinning in the same direction, the teeth will smash into each other. But if the two gears are spinning in opposite directions, the teeth can mesh. The wires are pulled toward each other.

9.2 Maxwell-Faraday Law

How can we explain the way in which a changing magnetic field can generate an electric field: $\nabla \times \boldsymbol{E} = -\frac{1}{c}{\partial \boldsymbol{B}}/{\partial t}$? When we attach a battery to a circuit, the wire becomes an electric field that frees the lightly held valence electrons that are present in conductors. The freed electrons then move in the field due to electrical attraction. In the case of an electric field produced by induction, the electrons must be made to move without a source of electrical attraction. Since the motion of the electrons in the induced current described by the Maxwell-Faraday Law is due to a magnetic field, we know that this force will be entirely mechanical.

If I move a conductor perpendicular to a magnetic field, I will generate a current in the conductor that is also perpendicular to the magnetic field. How is this possible? Using Cartesian coordinates, we can imagine that the wire is along the z axis and the motion of the wire is in the y direction. The direction of the magnetic field is in the x direction, and the current is generated in the z direction along the wire. The magnetic field forces the weak-force forcelines of the electrons to align parallel with the field, so that the weak-force forcelines lie in the plane created by the x and z axes. This will be explained further in chapters 13, 14, and 15. The electrostatic forceline of the electron must now point along the wire in the z direction because the electrostatic forceline of the electron is always perpendicular to the weak-force forcelines of the electron.

The spinning w-duals of a magnetic field can be considered as being like paddlewheels. Like the push of a paddlewheel, the direction of the push of a w-dual flips from top to bottom. If the wire simply rests in the magnetic field, the weak-force forcelines of the electrons will be hit roughly equally in either direction along the z axis. If I now raise the wire in the magnetic field, the electrons will come into contact with the bottom of the w-dual paddlewheels more often than the top and feel more of an impetus in that direction than if the electrons were hit more often by the top of the paddlewheels. The faster I raise the wire, the more w-duals of the magnetic field that the weak-force forcelines of the electron will hit and the greater the voltage. Lowering the wire will send electrons in the opposite direction. Reversing the magnetic field will also reverse the direction of the electron flow.

We can now test whether our decision to regard counterclockwise-spinning w-duals (when looked at from behind) as pointing in the north direction of the magnetic field was correct. Using the left-hand version (flow of electrons rather than current) of Fleming's right-hand rule, where our left thumb points in the direction that the wire is moving (y), our index finger in the direction of the magnetic field (x), and our middle finger sticking out perpendicular to our palm in the direction that the electrons are flowing (z), moving the wire upward in the direction of the thumb (y) should generate an electron flow out of our middle finger (z). The direction of our index finger is toward the north pole of the magnetic field, so that the w-duals must be spinning counter-clockwise. Moving the wire upward brings the electrons in the wire into contact with the bottom of the spinning w-duals more often than the top, pushing the electrons toward the end of our middle finger. Reversing the movement of the wire brings the electrons into contact with the top of the spinning w-duals more than the bottom, reversing the direction of the flow of electrons. This only works if the w-duals are spinning in a counterclockwise direction when looked at in the direction that our index finger is pointing. The decision about the spin of the w-duals in a magnetic field in the last chapter was not arbitrary.

9.3 Empty Space Solution

In empty space, there are no charged particles, and the four Maxwell equations due to Heaviside can be simplified. I will use Gaussian units:

$$\nabla \cdot \boldsymbol{E} = 0$$
$$\nabla \cdot \boldsymbol{B} = 0$$
$$\nabla \times \boldsymbol{E} = -\frac{1}{c} {}^{\partial \boldsymbol{B}}\!/_{\partial t}$$
$$\nabla \times \boldsymbol{B} = \frac{1}{c} {}^{\partial \boldsymbol{E}}\!/_{\partial t}$$

Starting with the fourth equation, we can take the curl of both sides, which gives:

$$\nabla(\nabla \cdot \boldsymbol{B}) - \nabla^2 \boldsymbol{B} = \frac{1}{c}\frac{1}{\partial t}(\nabla \times \boldsymbol{E})$$

Div **B** equals 0, and I can substitute for the curl of **E** using the third equation. I will then have:

$$\nabla^2 \boldsymbol{B} = \frac{1}{c^2}\frac{\partial^2 \boldsymbol{B}}{\partial t^2}$$

I can do the same thing for the electric field and will end up with the same equation, but for the electric field this time. These equations are in exactly the same form as the wave equation, and suggest that it is possible to create either a magnetic or an electric wave that travels through space at the speed of light. This is an embarrassment of riches because we have never seen separate magnetic and electric waves in empty space.

The solution to the problem is an electromagnetic wave that combines both. This is not possible in SUiTe. There is no means by which an electric wave can propagate in space without electrons. The magnetic wave cannot induce an electric field without electrons. So in SUiTe, only the wave equation regarding an alternating magnetic field is correct.

A radio broadcast antenna has an alternating current overladen with all the necessary information for a radio to turn the subsequent induced current into sound. The wave generated by the radio wave source is a perfect wave, like the wave of water generated in a pond if you were to drop a pebble in the center. The wave of water races out from the center, and the amplitude decreases as the circular wave grows wider. The total energy of the wave remains the same.

The wave that the radio antenna receives that creates a current has a small fraction of the energy that created the original wave. We can imagine a wave train stretching from our radio's antenna all the way to the broadcast source. In chapter 10, on photons, we will find that this beautiful and accurate picture of radio waves emitted by an alternating current has almost nothing to do with a photon emitted by an atom. The confluence of the radio wave of Maxwell and Hertz and the photon has obscured an important truth of our universe. The photon is a pulse. Einstein's description of the photon as a packet of energy is the most accurate description.

Chapter 10

Photons

After a series of experiments, Newton concluded that light was a particle. For over one hundred years, until the dawn of the nineteenth century, light was treated as a particle. Then Young passed light through two closely spaced slits in a barrier and observed an interference pattern some distance from the barrier. There were light and dark bands, places that light avoided and places where it accumulated. Similar interference patterns had been seen using water waves. Additional work was done, by Young and others, to show that in every case where Newton had experimental evidence for light as a particle, a wave theory for light would have worked equally well, even if, at times, the resulting wave theory was more cumbersome. Waves could explain all known phenomena, and particles couldn't. Light was a wave.

This success lasted almost exactly 100 years until the first year of the twentieth century, when Planck discovered that the only way to model blackbody radiation was to assume that light came in packets of energy. The energy of a wave depends on two values—its amplitude and its frequency. The frequency is the number of waves per unit of time. The amplitude is the height of the wave. Planck's model excluded the amplitude. Light was now a very odd wave. It was better to describe it as a particle with some wave-like properties. Einstein, in 1905, solved the photoelectric effect problem using a new constant based upon Planck's theory that we now call h, and the frequency of the light radiation that was sufficient to dislodge an electron from a metal surface. Eighteen years later, Compton, while studying the momentum of light, found that gamma-ray particles obey the conservation of momentum when they strike electrons. The electron that was struck suddenly accelerated. And the photon? It can't speed up or slow down, so it changed its energy in the only way possible, by changing its frequency. This was again a result that was not wave-like.

The wave theory of light had successes after Compton for the rest of the 1920s. De Broglie theorized that matter had wave properties in 1924 and was proven correct. Schroedinger published a wave equation that provided a justification for the choices Bohr and others had made regarding the hydrogen atom, and Heisenberg, in 1927, explained his uncertainty principle in terms of a perfect microscope that was able to use light of any frequency in order to determine the exact location of a particle. This description depends upon the diffraction limit of light, a wave property. Photons, like electrons, have been placed in the box of wave-particles. If our experiment is looking for a particle, we will find one. If it is looking for a wave, we will find one. In some experiments, an electron, or a photon, can manage to switch between particle and wave.

10.1 A Photon Is a Phonon

The name *phonon* was partially chosen to match the name photon. The study of phonons, wave-particles of sound moving through a crystal, has shown a number of similarities with the behavior of photons. As a phonon moves through a crystal, it compresses the crystal in the longitudinal direction (the direction of motion), but it also creates a transverse compression (a compression perpendicular to the direction of motion). It is the transverse compression that sketches out the wave that we are used to for light. The transverse compression begins at the beginning of the longitudinal compression as it moves through the crystal at the speed of sound for that crystal. The transverse compression expands at the speed of sound, reaching its maximum where the longitudinal compression is also at its maximum.

As the phonon passes through some given location, the longitudinal compression first increases than decreases before disappearing. The transverse compression begins as the longitudinal compression begins, spreading at the speed of sound. It reaches its maximum distance where the longitudinal wave is most compressed, and then begins to collapse, decreasing as the longitudinal wave decreases, ultimately disappearing when the longitudinal wave disappears. One can imagine a sound pulse passing through a crystal as being similar to a tire swallowed the wrong way passing through a python. The diameter of the tire is one wavelength and is perpendicular to the direction of motion. The thickness of the tire is one-half wavelength.

Over distance, phonons spread, and more of the nodes of the crystal become involved in the pulse. The transverse wavelength then grows larger, and the phonon becomes less compressed in the longitudinal direction. The degree of spreading depends on the wavelength in comparison to the distance between nodes in the crystal. If the wavelength is large relative to the internode distance, there is very little spreading. If this is also true for a photon, then extremely powerful gamma-ray photons, those that have very small wavelengths, won't be able to travel long distances—billions of light-years—without spreading. There will be a cutoff point for gamma-ray photons that is associated with their energy and distance. There will also be a limit to how small a wavelength of light can be, and how much energy a photon can have. In SUiTe, space behaves like an ideal crystal because of the enormous pressure space is under, and the repulsion that the duals of space have for each other. If we could see space as it is, it would look like two, interleaved, three-dimensional lattices.

We have measured very high-energy gamma rays on Earth, which suggests that an analogy between photons and phonons shouldn't go too far. It is nevertheless possible that many of these energetic gamma-ray photons will turn out to be particles of dark matter. The only way to separate the two is to determine that the photon, after a collision, is still moving at the speed of light, though with a larger wavelength. If there is any slowdown, the collision was caused by a particle of dark matter.

Whether or not parts of the analogy with the phonon fail, the photon is also a pulse with both a longitudinal compression and a transverse compression. Most light is the result of the bremsstrahlung of an electron. In the case of a plane wave, the resulting longitudinal compression is not disk-like but more in the form of a rectangular solid. The resulting transverse compression is not disk-like either, but also rectangular. The longitudinal compression of the photon is responsible for the photon's momentum, while the transverse compression provides the necessary information for scattering, interference, and deflection. Massive particles also create a longitudinal compression of space in front of them. The scattering of an electron by a particle of dark matter should be nearly identical to the scattering of an electron by a gamma-ray photon. In both cases, it is the compression of space between the two particles as they come close to each other which causes the momentum transfer.

10.2 The Quantum Spin of a Photon

The most common way in which photons are created is by the sudden deceleration of an electron. Experiments with light use those sorts of photons. Because the transverse compression is on the plane that is perpendicular to motion, and it is rectangular in shape, it is possible to think of the transverse compression as being like a bundle of rods moving through space. The lengthwise direction of the bundle is perpendicular to the direction of motion. When sunlight reflects off water, some of the photons are polarized, meaning that the transverse waves of those photons are set to the same angle as the surface of the water. Polarizing sunglasses absorb those photons but let the others pass through. The polarization of the photon is different than the polarization of the electron. The longitudinal compression of the photon is a bundle of compressed w-duals. A 180-degree rotation of that bundle results in an identical bundle. A 180-degree rotation of an electron in a Stern-Gerlach apparatus flips the spin. This is different. There are three possible spin states for the electron, but only two for the photon.

Two electrons moving parallel with each other must rotate, so that their spins are perpendicular. When two electrons can maintain this state, it is a Cooper pair. Photons have no mass and cannot generate a rotation in each other. Like waves at the ocean, they can move through each other, though when they move through each other they are able to combine their energy. Since all photons move at the speed of light, they cannot catch up to each other like waves on the ocean. Two photons must be at an angle with respect to each other in order to temporarily join. By adjusting the angle of two lasers, it is possible to choose the location at which photons coming from the two lasers can potentially join. This property of light has been used in two-laser microscopy to excite the interior of a material with the same energy that a higher energy photon would have, when the higher energy photon would not have been able to move through the material toward the interior without having been absorbed prior to reaching the location of interest.

Photons created by the deceleration of electrons are always plane waves, but two plane waves can be very close to each other if their w-dual bundles are perpendicular. These combinations of perpendicular plane waves can best be described as circularly polarized waves. A circularly polarized beam of light can be split into its two perpendicular components by some substances.

When discussing spin in the next chapters, I will confine the theory to electrons, even though photons are used almost exclusively in tests of quantum entanglement and of Bell's inequalities. The reason for avoiding photons is that there is no clear explanation of what happens to a photon when it interacts with a substance such that its polarity changes. Is the photon absorbed and re-emitted? This is my preferred explanation, but it is possible that the atom or molecule can somehow modify space in such a way that the polarity of the photon is changed. Light that is slowed via the Shapiro effect as it passes close to the Sun is not absorbed and re-emitted. It is space that has been modified. The question of the photon passing through substances that change its polarity is ultimately a question for experimenters. Light does slow when passing through materials, but is there the possibility of making an extremely thin slice of some substance that manages to change the polarity of a photon without reducing the speed of the photon? If that is possible, then the photon is not absorbed and re-emitted.

Chapter 11

Kaon Branching Fractions

I would like to step away from the chronological development that led from Maxwell's equations to quantum electrodynamics (QED). Although SUiTe reproduces classical physics and special and general relativity, it does not duplicate the standard model, especially particle physics. The branching fractions of the kaon cannot be accurately calculated using the standard model because of the difficulty of modeling quark wave functions. If the wave function could be determined for each of the branches, it should be possible to use Fermi's Golden Rule for decays in order to find the branching fractions. Fermi's Golden Rule for decays depends on the mass and momentum of the decaying particle, the masses and momenta of the particles created in the particular decay branch chosen, and the wave function. I will show that all the branching fractions for the kaon can be theoretically determined without the wave function or the outgoing masses and momenta.

The presently accepted path for computing the branching fractions is to determine the decay rate for a particular decay channel; then add all the rates for a specific particle—the charged kaon, for example—to obtain the total decay rate; and, finally, divide the channel decay rate by the total decay rate. The result is the decay branching fraction for that specific particle decaying through that particular channel. Due to the difficulty of calculating the wave functions for the quarks from which kaons and other hadrons are believed to be made, it has not been possible to use Fermi's Golden Rule except in the simplest of particle decays. As a result, the method has not been tested for mesons larger than pions.

The three smallest particles that decay are the muon, the pion, and the kaon. The muon is not a hadron and is therefore not constructed of quarks. It can only decay into an electron or a positron along with neutrinos because there are no other charged particles larger than electrons and positrons and smaller than muons. The muon decays using the weak force. Its decay is used to determine the coupling constant of the weak force. Since the coupling constant is also necessary for finding the wave function of pion and kaon decays, the decay of the muon is not an adequate test of the Golden Rule by itself. The decay of the muon in combination with the decays of the pion or the kaon, however, could show that the Golden Rule for decays works.

The pion decays into either a muon or an electron along with neutrinos. For the pion, the decay is strongly limited, and the two branching fractions can be determined without finding the wave function. Here again, the Golden Rule is not used. The kaon, then, represents the first major test of Fermi's Golden Rule for decays.

Kaon Branching Fractions

Particle	Decay Products	Measured Value*	Estimate
$K^{\pm}$	$\mu^+ \nu_\mu$	63.56 $\pm$.11	63.58
	$\pi^0 e^+ \nu_e$	5.07 $\pm$.04	5.05
	$\pi^0 \mu^+ \nu_\mu$	3.352 $\pm$.033	3.368
	$\pi^+\pi^0$	20.67 $\pm$.08	20.67
	$\pi^+\pi^0\pi^0$	1.760 $\pm$.023	1.756
	$\pi^+\pi^+\pi^-$	5.583 $\pm$.024	5.578
K_S^0	$\pi^0\pi^0$	30.69 $\pm$.05	30.65
	$\pi^+\pi^-$	69.20 $\pm$.05	69.24
K_L^0	$\pi^{\pm} e^{\pm} \nu_e$	40.55 $\pm$.11	40.57
	$\pi^{\pm} \mu^{\pm} \nu_\mu$	27.04 $\pm$.07	27.05
	$\pi^0\pi^0\pi^0$	19.52 $\pm$.12	19.50
	$\pi^+\pi^-\pi^0$	12.54 $\pm$.05	12.53

* The measured values were taken from the particle listings of P.A. Zyla et al. (Particle Data Group), Prog. Theor. Phys. 2020, 083C01 (2020) and 2021 update.

To have 12 estimates, including all the major branching fractions of the kaon, within the error margins of the measured data has never been done before in physics. This shouldn't be possible, especially since it does not follow the Golden Rule, which is a reasonable rule if the underlying description of matter is as imagined. Apparently, it isn't.

In order to find an organizing principle that allows us to see the connections between the various decay branches without a corresponding wave function, it was necessary to look for patterns in the data that would suggest formulas that could be used to find the branching fractions. Such patterns do exist and suggest that nature does not follow the Golden Rule, nor does nature follow any other path that our intuition would have suggested.

For example, there would be no reason to suspect that the long-lived neutral kaon would have any effect on the branching fractions of a charged kaon. Charged kaons and long-lived neutral kaons are different particles with different masses and mean lifetimes. Nature confounds our expectations by determining all the branching fractions together, using a single model that is only dependent on the mass ratio of neutral kaons to charged kaons. I will use the symbol $K^{\pm}$, to refer to both positive and negative charged kaons, K_L^0 for long-lived neutral kaons, and K_S^0 to refer to short-lived neutral kaons. The previous table gives the measured and estimated values for the 12 major branching fractions of the kaon.

11.1 Pairs of Pions and the Construction of Kaons

The model begins with an assumption about the relative numbers of the different types of pions that nature has available when constructing kaons. In this model there are four pions: the two charged pions, π^+ and π^-, the neutral pion, π^0, and a slightly different neutral pion which decays directly into dark matter and cannot be detected, the dark-matter pion, π_f^0. In section 20.6, where I attempt to determine the mass and properties of the dark matter particle, I suggest using the name fanton for dark matter. This is the reason for using the subscript f in π_f^0 for the dark-matter pion. When I refer to a neutral pion, it is the normal neutral pion. I will refer to the other, much more common neutral pion as the dark-matter pion. Charged pions, including both positive and negative pions, are as common as normal neutral pions. Dark-matter pions are nine times as abundant as either charged or neutral pions.

Every kaon appears to be made from three pions and one quasi-particle constructed from neutrinos. What I will call the *apex particle* is the one opposite the quasi-particle. Kaons are in the shape of a square. In SUiTe, the alpha particle is also laid out flat in the shape of a square. This allows the alpha particle to share spin as well as charge with protons opposite each other and neutrons opposite each other. For the mesons, the square shape provides them with a stability from spin sharing that is essential for modeling the kaon. I will not indicate whether a charged pion is positive or negative. If there are two or three charged pions in a kaon, two of them must have opposite charges.

Treating both negative and positive pions as a single charged pion, there are only three different types of pions. With three different pions and one quasi-particle, there could be as many as 18 different kaons, but certain combinations are not allowed by nature. Dark-matter pions cannot be adjacent to each other. If there are two or more neutral pions in a kaon, two must be opposite each other. A neutral pion cannot be opposite a charged pion. This leaves 11 possible kaons, of which 10 have been detected. The last entry on the left in the following table should be the most common kaon according to this model. It is constructed from two dark-matter pions and one neutral pion, with the neutral pion at the apex. Because it decays into two particles of dark matter and a single neutral pion, it would be difficult to determine that the neutral pion had previously been part of a kaon. Although this kaon has never been measured, it is an essential part of this model and plays a major role in the construction of tau particles and D mesons.

To create a kaon, it is necessary to start with a pair of pions. Given the relative abundance of charged, neutral, and dark-matter pions, if $\pi^{\pm}\pi^{\pm}$ and $\pi^0\pi^0$ represent one pair each, there will be 39 other pairs for each $\pi^0\pi^0$ or for each $\pi^{\pm}\pi^{\pm}$ that is made. Two out of 40 pairs will be $\pi^{\pm}\pi^{0}$ pairs because, in this pair, either the neutral pion or the charged pion could be the eventual apex particle of the resulting kaon. There will be 18 $\pi^{\pm}\pi_f^0$ pairs because the dark-matter pion, π_f^0, is 9 times more common than the charged or neutral pion, and either particle could become the apex particle. The same is true for the $\pi^0\pi_f^0$. Although the following model can be developed directly from the 40 pairs and subsequently add up to 40, the form of the equations is simpler if we multiply through by 4 and have the model add up to 160.

At a total of 160, the probability that nature will make a $\pi^0\pi^0\pi^0$ kaon will be $4/160$ because there is only one kaon that can be constructed starting from a neutral pion pair. The same is not true of the $\pi^{\pm}\pi^{\pm}$ pair, which can add either a dark-matter pion or a charged pion. The $\pi^{\pm}\pi^{\pm}\pi^{\pm}$ kaon will be $4/160$ minus a contribution to the $\pi^{\pm}\pi^{\pm}\pi_f^0$ kaon in the final model. This contribution is the result of a different decay rate for the charged kaon and the long-lived neutral kaon and is not found in the pure model.

The model is exponential in a single variable x. There is a zeroth element whose value is 1. The kaon associated with the zeroth element is $\pi^0\pi^{\pm}\pi^0$. When moving from the pure model, which adds up to 160, to the actual model, which adds up to more than 160, the values of the $\pi^0\pi^0\pi^0$ kaon and the $\pi^0\pi^{\pm}\pi^0$ kaon are not affected, as can be seen by comparing the formulas from the pure model to the actual model. The $\pi^{\pm}\pi^{\pm}\pi^{\pm}$ kaon is only affected indirectly through the contribution that it makes to the $\pi^{\pm}\pi^{\pm}\pi_f^0$ kaon. Whenever I refer to a kaon by its pion components, the center particle is the apex particle.

In the pure model, we will assume that adjacent connections between pions in the kaon have the same probability, that the masses of all kaons are equal, and that the mean lifetimes of all kaons are equal. The $\pi^0\pi^0\pi^0$ kaon and the $\pi^{\pm}\pi^{\pm}\pi^{\pm}$ kaon are not found in the following table because both have been set to 4 independent of x. The pions in the brackets are not part of the formula but are there to indicate which kaon the formula corresponds to. The problem is to solve for x such that all the terms plus 8 add up to 160.

11.2 The Pure Model

$x^0\ [\pi^0\pi^\pm\pi^0]$		
$x^1\ [\pi^\pm\pi^0\pi^\pm]$		
$x^2\ [\pi^0\pi^0_f\pi^0]$		
$x^3\ [\pi^\pm\pi^0_f\pi^\pm]\ 8/9$		
$x^4\ [\pi^0_f\pi^\pm\pi^0_f]\ 8/9$	$1/3\ x^4\ [\pi^\pm\pi^\pm\pi^0_f]$	$1/9\ x^4\ [\pi^0\pi^\pm\pi^0_f]$
$x^5\ [\pi^0_f\pi^0\pi^0_f]\ 2/3$		$1/9\ x^5\ [\pi^\pm\pi^0\pi^0_f]$

The bottom left term corresponds to the kaon that has not yet been measured. As there is no measured branching fraction for this kaon, we will only include it in the pure model, but not when we solve the model for those decays whose values have been measured. The value for x for which the formulas in the above table plus 8 add up to 160 is approximately 2.501251.

These formulas for 9 of the 11 kaons that lead to the branching decay ratios (the other two, $\pi^0\pi^0\pi^0$ and $\pi^\pm\pi^\pm\pi^\pm$ are exactly equal to 4 in the pure model independent of x) came from examining the measured values of the branching ratios and noting that there was an exponential relationship between them. The resulting relationship was simple as can be seen from the table. There is no theoretical reason at this time from SUiTe why this should have been the case, even though the structure of the kaons is provided by the theory. Another important piece of data necessary for developing the model was the relative abundance of each of the three types of pions. Charged and neutral pions were known to have the same abundance, but the abundance of the dark-matter pion could only be determined by noting the importance of beginning the process of constructing kaons with 40 possible pairs of pions.

11.3 Modeling Kaons with Pion Pairs

Here is a schematic indicating the ways in which the five unique pairs among 40 possible create the 11 kaon structures. Some combinations are repeated. Each kaon requires three pions.

Kaon Structure from Pion-Pairs Schematic

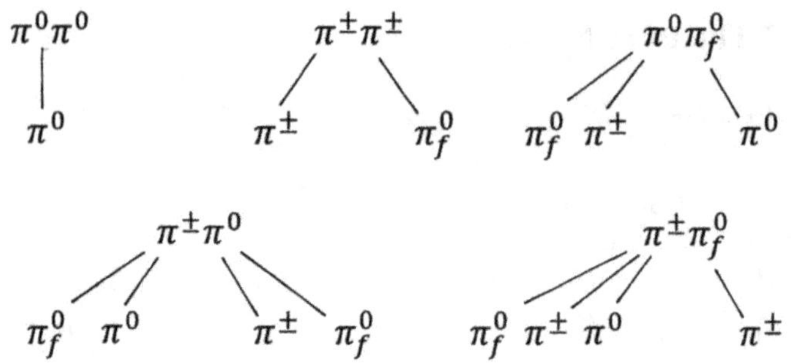

11.4 Kaon Structure

Kaons are constructed from three pions plus a quasi-particle made from neutrinos. The kaon can be imagined in the form of a square. The 11 possible kaons, in order from the smallest value to the largest as given by the model formulas, are:

	$\pi^\pm$			π^0			$\pi^\pm$			π^0	
π^0		π^0	$\pi^\pm$		$\pi^\pm$	$\pi^\pm$		$\pi^\pm$	π^0		π^0
	ν			ν			ν			ν	
	$\pi^\pm$			π_f^0			π^0			$\pi^\pm$	
π^0		π_f^0	π^0		π^0	$\pi^\pm$		π_f^0	$\pi^\pm$		π_f^0
	ν			ν			ν			ν	
	π_f^0			$\pi^\pm$			π^0				
$\pi^\pm$		$\pi^\pm$	π_f^0		π_f^0	π_f^0		π_f^0			
	ν			ν			ν				

11.5 Qualitative Analysis of the Kaon Structure

The set of possible kaon structures given above can be separated into two groups based upon whether they are charged or neutral. The group of neutral kaons can then be split into two groups, noting that only two kaons place the π_f^0 at the apex. This gives us three different groups to match to the three different mean lifetimes. Those neutral kaons with π_f^0 as the apex particle decay much more rapidly than those without. I have not been able to find a way to estimate the ratio of the mean lifetimes from the branching ratios of the two groups of neutral kaons because there are only two kaons where there is a single π_f^0 that appears in the apex, and those two kaons do not contribute to the branching fraction of either a charged or a long-lived neutral kaon.

When two dark-matter pions are adjacent to a charged pion at the apex, the kaon decays exclusively into a muon and muon neutrino. When the structure includes a single dark-matter pion and it is adjacent to a charged pion that is at the apex of the kaon, the charged pion decays either into an electron plus an electron neutrino or into a muon plus a muon neutrino with a strict 3:2 ratio. If there is only one dark-matter pion in the structure and it is at the apex, the decay is rapid and does not bring about an immediate decay in the other pions.

11.6 The Actual Model

It is necessary to modify the pure model given that some adjacent bonds are more probable than others, and that the masses and mean lifetimes are not equal for all kaons. We start by multiplying x from the pure formula by the mass ratio. From the 2021 *Review of Particle Physics* summary tables, the mass ratio of the long lived neutral kaon to the charged kaon, R_m, is approximately 1.007969. After multiplication from the solution for x from the pure formula, we have 2.5212 for x. This affects all nine formulas equally as it should. In this model, the probability of the creation of a particular kaon ultimately determines the probability of a particular decay path since each decay path follows from a particular kaon. The probability of making a particular kaon depends partially on the mass ratio.

The next step is to multiply each term that includes the pair $\pi^{\pm}\pi_f^0$ by the mass ratio, indicating that this bond is more likely than other bonds. The bond that links the pair $\pi^{\pm}\pi^0$ is slightly more probable than the bond for either $\pi^{\pm}\pi^{\pm}$ or $\pi^0\pi^0$. In this case, the adjustment is not by a simple multiplication by the mass ratio; instead, we have a column of four kaons with a multiplier that goes from 0 to 3. The number that it multiplies can be determined by first noting that the probability of a $\pi^+\pi^0$ pair is $1/20$ in the pure model. We then divide the pair probability by *x* to get the adjustment value, which is 0.01983.

$\pi^0\pi^\pm\pi^0$	$1 + 0 \cdot 0.01983$
$\pi^\pm\pi^0\pi^\pm$	$1 + 1 \cdot 0.01983$
$\pi^\pm\pi^0\pi_f^0$	$1 + 2 \cdot 0.01983$
$\pi^0\pi^\pm\pi_f^0$	$1 + 3 \cdot 0.01983$

The last step is to determine how much of the $\pi^\pm\pi^\pm\pi^\pm$ kaon probability is associated with the $\pi^\pm\pi^\pm\pi_f^0$ kaon. The amount that must be subtracted from the $\pi^\pm\pi^\pm\pi^\pm$ kaon is $z_\pm$. This value will have to be adjusted in order to be added to the $\pi^\pm\pi^\pm\pi_f^0$ kaon because the $\pi^\pm\pi^\pm\pi^\pm$ kaon is in the charged group and the $\pi^\pm\pi^\pm\pi_f^0$ kaon is in the long-lived neutral group. The number added to the $\pi^\pm\pi^\pm\pi_f^0$ kaon will be z_L. We can find $z_\pm$ and z_L. from the equation:

$$z_\pm \cdot z_L = R_m \cdot \Gamma_L^0/\Gamma^\pm$$

where the last term is the ratio of the total decay constants of long-lived neutral kaons to charged kaons and R_m is the mass ratio. We also have the equation:

$$z_\pm = z_L \cdot \sum K^\pm / \sum K_L^0$$

where the sums are the sum of the terms from the model for charged kaons and long-lived neutral kaons, respectively. There is also an equation to determine the ratio of the decay constants for long-lived neutral kaons and charged kaons.

$$\Gamma_L^0/\Gamma^\pm = 2/3 \cdot R_m \cdot 0.9965^{-1} \sum K_L^0 / \sum K^\pm = 0.2421$$

The additional factor of 0.9965^{-1} is necessary because the four major branching ratios of K_L^0add up to less than 1.

$z_{\pm} = 0.8244$ and $z_L = 0.2960$. There is one last detail: for both the $\pi^0\pi^\pm\pi_f^0$ kaon and the $\pi^\pm\pi^\pm\pi_f^0$ kaon, nature splits the decay products in the ratio of 3:2, where 60 percent of the time the decay includes an electron and an electron neutrino and 40 percent of the time it includes a muon and a muon neutrino. We can now write out the equations for the various decay channels. The equations are shown in the following table.

Branching Fraction Formula Tables

Decay Products	Formula	Formula Value	Fraction Estimate*
$\boldsymbol{K^\pm}$			
$\mu^\pm\ \nu_\mu$	$x^4R\cdot 8/9$	36.201	63.58
$\pi^0\ e^\pm\ \nu_e$	$x^4R\cdot 0.6/9\cdot 1.0595$	2.877	5.05
$\pi^0\ \mu^\pm\ \nu_\mu$	$x^4R\cdot 0.4/9\cdot 1.0595$	1.918	3.368
$\pi^\pm\pi^0$	$x^5\cdot 1/9\cdot 1.0397$	11.768	20.67
$\pi^\pm\pi^0\pi^0$	x^0	1	1.756
$\pi^+\pi^+\pi^-$	$4 - z_\pm$	3.176	5.578
		$\sum K^\pm = 56.940$	
$\boldsymbol{K_S^0}$			
$\pi^0\pi^0$	x^2	6.356	30.65
$\pi^+\pi^-$	$x^3R\cdot 8/9$	14.359	69.24
		$\sum K_S^0 = 20.715$	
$\boldsymbol{K_L^0}$			
$\pi^\pm\ e^\pm\ \nu_e$	$(x^4/3 + z_L)R\cdot 0.6$	8.324	40.57
$\pi^\pm\ \mu^\pm\ \nu_\mu$	$(x^4/3 + z_L)R\cdot 0.4$	5.549	27.05
$\pi^+\pi^-\pi^0$	$x\cdot 1.01983$	2.571	12.53
$\pi^0\pi^0\pi^0$	4	4	19.50
		$\sum K_L^0 = 20.444$	

* The estimates for K_S^0 were multiplied by 0.9989, the total that these two branching fractions represent in the data given in the

2021 *Review of Particle Physics* summary tables. The estimates for K_L^0 were multiplied by 0.9965.

11.7 Conclusion

The remarkable accuracy that this model provides for estimating the branching fractions of the kaon suggests that the decays depend on the possible kaon structures that can be made from the two charged pions and the two neutral pions of this theory, and how probable the creation of each kaon is. The mass ratio appears in the model as a correction to a pure model. Knowledge of the decay products, their momenta, and the wave function of the decaying particle were unnecessary in determining the branching ratios. This suggests that Fermi's Golden Rule for the decay of particles such as mesons is unnecessary and that it does not describe the process that nature follows.

It is a surprise that the ratio of the total decay rates for the charged and long-lived neutral kaons depends only on the model and the mass ratio. The linking of charged and long-lived neutral decays, through the branches that require $z_\pm$ and z_L, makes this possible. Because of this linkage, the knowledge of a single decay rate would allow 10 of the 12 decay rates as well as 2 of the 3 total decay rates to be determined. There is no similar link to the short-lived neutral kaon.

Unlike in the standard model, the likelihood of creating the particular charged or neutral kaon from the various pions that nature has available determines the decay branch. The most common pion that can be created is the dark-matter pion. The most common decay path for a charged kaon, the one which decays into a muon and a muon neutrino, also includes two dark-matter pions. It represents 63.58 percent of charged kaon decays.

The pure model (section 11.2) can be solved for the value of x which is then tweaked using the mass ratio of the kaons. What about a particle like the tau which has a single mass? We will see in the next chapter that the pure model value for x from kaons is the correct value for the equation of the tau branching decay ratios.

Chapter 12

Tau and D Meson Branching Fractions

In the previous chapter, it was possible to determine all possible kaons and their decay channels directly from the theory. If the neutral and charged kaons had had the same mass, it would have been possible to determine the 12 branching decay fractions directly from the theory without making any adjustments. This would have been the first time that a quantitative result within measurement error had been done from theory alone within physics. Although requiring only the ratio of the masses of the neutral and charged kaons was close to perfect, it did not meet the goal that theoretical physicists have had for more than 100 years of determining a quantitative result directly from theory. The two most popular targets for theoretical physicists have been the fine structure constant and the proton-electron mass ratio. No one, to my knowledge, has considered that probabilities, such as those of the branching decay ratios of temporary particles, are also dimensionless.

Both the kaon and the tau have been carefully studied. The present range of errors in the measurement of the branching decay ratios is unlikely to change because they are now similar to the other observational errors for these two particles such as mass and mean lifetime. The theoretical estimation of the branching decay ratios of the kaon in the previous chapter made clear the dependence on the measurement of the masses of the two types of kaons. The remaining inaccuracy of their two masses is explained by SUiTe as the result of small, but important, changes in the masses of different versions of the charged or neutral kaons. Each version must be weighed separately. The same is true of the tau, but it is even more difficult because the same decay channel usually has at least two potential precursors. Fortunately, when a path is shared, the ratio of their decays is a simple fraction. The tau particle, because the masses of the various versions of the tau are very nearly the same, is the ultimate test of the "Pure Model" developed for the kaon in section 11.2.

12.1 Tau Lepton

The advantage of studying kaons is that they provide a complete list of the possible combinations of three pions. For all larger particles, there is competition between particles of roughly the same mass for those kaons that can be made. The fanton kaon branching decay ratio has never been measured though it can be determined from theory. It is a kaon constructed of two fanton pions opposite each other with a neutral pion at the apex. Because the fanton kaon does not use the charged pion, free neutral pions become less common than charged pions, and the kaon made from three neutral pions disappears entirely in the tau decays. All tau particles have two fanton kaons opposite each other, so all tau particles depend on the apex kaon and the particle opposite for their probability. This simplifies the determination of the branching decay probabilities. Either the apex kaon or the particle opposite must be neutral while the other must be charged. Since charge is a binary property, the divisions of coefficients by ninths in the pure model of section 11.2 that describes a single kaon will change to eighteenths in this model.

A difficulty with the tau is that the same decay branch will usually have two or even three possible beginning structures. One in particular, that decays into $3\pi^{\pm}$, has three possible starting structures: K_S^0 with $\pi^{\pm}$ opposite each other; $K^{\pm}, \pi_f^0$; and $K_L^0, \pi^{\pm}$. Fortunately the coefficients must remain simple. Only one branching decay ratio of the tau refers back to the pair probabilities, from which the pure model of the kaons are drawn, which are in fortieths. This is a major decay which should have been one to one, but because of the shortage of neutral pions becomes 41:40. As always in decays, these ratios are not approximate, but they are exact.

The polynomial for the tau decay has 8 terms, but ten different decay fractions. This is one more term than in the kaon model (section 11.2), but the extra term is a slight adjustment to the x^4 term which has been separated out partly for simplifying the expression and partly to draw attention to it. As before with the kaon with ten separate equations for the different kaons but 12 decay fractions, there are only seven separate equations that create nine decay fractions.

The five basic structures for the tau are

	$K^0_{L,S}$			K^0_{Lf}			$K^\pm$			$K^\pm$	
K^0_{Lf}		K^0_{Lf}	K^0_{Lf}		K^0_{Lf}	K^0_{Lf}		K^0_{Lf}	K^0_{Lf}		K^0_{Lf}
	$\pi^\pm$			$\pi^\pm$			π^0_f			π^0	

The first represents two different tau structures with two different neutral kaons at the apex. Four of six possible decays of neutral kaons are included in the two. The second structure is separated from the first because there are no measureable decays of this apex kaon and it is therefore a phantom, hence the name fanton kaon. All that is seen when this tau particle decays is a charged pion.

$${}^{4}/_{9}x^5 + {}^{17}/_{18}x^4 + {}^{1}/_{90}x^4 + {}^{4}/_{9}x^3 + {}^{x^2}/_{3} + {}^{15}/_{9}x^1 + x^0 + {}^{17}/_{18}4 = 98.90$$

The solution is $x = 2.501251$ the same solution as for the pure model of the kaon. The coefficients are different than in the pure model because they are affected by the particle that is opposite the apex kaon and the lack of neutral pions in comparison to the kaon. The two fanton kaons on the two sides of all tau structures allow tau particle probabilities to be determined by a single apex kaon and the particle opposite it. The equations for the 10 branching decay ratios are

${}^{8}/_{9}x^4 \cdot 0.5 = 17.40$	$\mu^\pm$	17.39 (.04)
${}^{8}/_{9}x^4 \cdot 0.5 \cdot {}^{41}/_{40} = 17.83$	$e^\pm$	17.82 (.04)
$({}^{1}/_{3}x^5 - {}^{1}/_{3}x^2)\,{}^{5}/_{6} = 25.48$	$\pi^\pm\pi^0$	25.49 (.09)
${}^{1}/_{9}x^5 = 10.88$	$\pi^\pm$	10.82 (.05)
$({}^{8}/_{9}4 + {}^{8}/_{9}x^1)\,{}^{4}/_{5} = 4.62$	$\pi^\pm\pi^\pm\pi^\pm\pi^0$	4.62 (.05)
$({}^{8}/_{9}4 + {}^{8}/_{9}x^1)\,{}^{1}/_{5} = 1.16$	$K^*(892)^\pm$	1.20 (.07)
${}^{7}/_{9}x^1 = 1.95$	$\omega\pi^\pm$	1.95 (.06)
${}^{1}/_{18}x^4 + {}^{4}/_{9}x^3 + {}^{1}/_{18}4 = 9.35$	$\pi^\pm\pi^\pm\pi^\pm$	9.31 (.05)
${}^{1}/_{18}x^5 + {}^{11}/_{18}x^2 = 9.26$	$\pi^\pm 2\pi^0$	9.26 (.10)
$x^0 = 1$	$\pi^\pm 3\pi^0$	1.04 (.07)

The measured values on the right in this table and the two following tables were taken from the Review of Particle Physics, P.A. Zyla et al. (Particle Data Group), Prog. Theor. Phys. 2020, 083C01 (2020) and 2021 update.

The breakdown of $K^{\pm}, \pi_f^0$ into either $\mu^{\pm}$ or $e^{\pm}$ is not exactly $1:1$ but $e^{\pm}$ is one fortieth more likely. One fortieth is the pion pair probability for either two charged or two neutral pions discovered in section 11.1.

What should be noted about these equations is that the connections between the kaons that are associated with the various powers of x in the previous chapter on kaon decays are the same as are found here. If a kaon is associated with x^3 in the previous chapter, it must also be associated with x^3 in the decays of the tau. The ratio of 41:40 which I have separated out as a ninetieth here after multiplication by the other factors, could also have been seen as the result of a $3:2$ ratio in the decay of kaons. Nothing in the equations for the tau is new. It follows the model found for the kaon. Two entries are interesting because they are intermediate results that decay very quickly. ${}^{15}/_{9}\, x^1$ stands out because the coefficient for x^1 was less than one for kaons. It includes an intermediate decay that involves an ω that represents ${}^{7}/_{9}\, x^1$ all by itself.

12.2 D Mesons

The D mesons are not much more massive than tau leptons and they have about the same mean lifetime. They are also constructed of three kaons and one pion. Unlike the tau lepton, they have only one fanton kaon, at the apex of the D meson. They then have two normal kaons instead of one. I haven't been able to determine how the probabilities of the two kaons interact in the D mesons. This section has a different purpose, to show the difference between a good result from theory and one from numerology. The equation for the tau is closely circumscribed by theory. The only adjustments are with the coefficients, and even those are limited to ninths and eighteenths. With the following charged D meson, the only limitation is that the exponent of x

cannot be more than 5. This is a good example of a successful application of numerology but not of physics.

The general structures of the charged D mesons are

$$\begin{array}{ccccccccccccc} & K_{Lf}^0 & & & K_{Lf}^0 & & & K_{Lf}^0 & & & K_{Lf}^0 & \\ K_S^0 & & K^\pm & K_L^0 & & K^\pm & K^\pm & & K^\pm & K_S^0 & & K_L^0 \\ & \pi_f^0 & & & \pi_f^0 & & & \pi^\pm & & & \pi^\pm & \end{array}$$

The general structures of the neutral D mesons are

$$\begin{array}{cccccccccccc} & K_{Lf}^0 & & & K_{Lf}^0 & & & K_{Lf}^0 & & & K_{Lf}^0 & \\ K_S^0 & & K_S^0 & K_S^0 & & K_L^0 & K^\pm & & K^\pm & K^\pm & & K_{L,S}^0 \\ & \pi_{f,0}^0 & & & \pi_f^0 & & & \pi_{f,0}^0 & & & \pi^\pm & \end{array}$$

These represent seven different structures for the neutral D meson and four for the charged D meson. With D meson decays, when there is a free π^0 and there are two free π_f^0, all three can join together to form a K_{Lf}^0, which will not be seen as part of the decay. For example, the observed charged D meson decay products of $K_S^0\pi^\pm = 1.562$ and $K_L^0\pi^\pm = 1.46$ both have the same $K^\pm = \pi^\pm\pi^0\pi_f^0$ and π_f^0 in their structures. They match the first two of the structures given above for the charged D meson. The $K^\pm = \pi^\pm\pi^0\pi_f^0$ and π_f^0 become $\pi^\pm$ and a K_{Lf}^0 of which only the $\pi^\pm$ is seen.

Charged D mesons

The sum of the chosen branching decay ratios for the charged D mesons is 66.10. The equation is

$$2/3\,x^5 + 4/9\,x^4 + 2x^3 + 7/9\,x^3 + 1/3\,x^2 + 1/3\,x^2 \cdot 1.035 = 66.10$$

The coefficient of 1.035 is the ratio of the K_S^0 to the K_L^0 from the Branching Fraction Formula Tables (section 11.6) found by dividing 14.359 by 13.873 where neither neutral kaon includes decays that contain π^0. $x = 2.1124$. The bolded value in the table below lies outside the error range.

$^1/_3 x^5\ ^5/_8 = 8.76$	$K^0 e^\pm$	8.73 (.10)
$^1/_3 x^5\ ^3/_8 = 5.26$	$K^*(892)^0 e^\pm$	5.40 (.10)
$^1/_3 x^5\ ^5/_8 = 8.76$	$K^0 \mu^\pm$	8.76 (.19)
$^1/_3 x^5\ ^3/_8 = 5.26$	$K^*(892)^0 \mu^\pm$	5.27 (.15)
$^2/_9 x^4 - ^1/_{18} x^4 \cdot 0.6 = 3.98$	$K^\pm \pi^\pm e^\pm$	4.02 (.18)
$^2/_9 x^4 - ^1/_{18} x^4 \cdot 0.4 = 3.76$	$K^\pm \pi^\pm \mu^\pm$	3.65 (.34)
$^1/_{18} x^4 = 1.11$	$2\pi^+\pi^-\pi^0$	1.16 (.08)
$x^3 = 9.43$	$K^- 2\pi^+$	9.38 (.16)
$^7/_9 x^3 = 7.33$	$K_S^0 \pi^\pm \pi^0$	7.36 (.21)
$^2/_3 x^3 = 6.28$	$K^- 2\pi^+ \pi^0$	6.25 (.18)
$^1/_3 x^3 = 3.14$	$K_S^0 2\pi^+ \pi^-$	3.10 (.09)
$^1/_3 x^2 \cdot 1.035 = 1.539$	$K_S^0 \pi^\pm$	1.562 (.031)
$^1/_3 x^2 = 1.49$	$K_L^0 \pi^\pm$	1.46 (.05)

The coefficients of the equation are excellent as are the estimates, but there is no relation in terms of the powers of x between the kaons composing the charged D meson and the equations for the kaon or tau decays. The value of x is different from that of the kaon or tau equations. Because both the power and the value of x are tied to a particular kaon decay, and because that is the only guide from theory that we have, abandoning both of them replaces theory with numerology.

Neutral D mesons

The sum of the chosen branching decay ratios for the neutral D mesons is 59.88. The symbol *R*, when included, is the ratio of the masses of the neutral and charged kaons. This has only been allowed because, oddly enough, it appears only for those values that were not modified when determining the kaon values. It is a reverse modification and perhaps has value. The equation is

$$0.2Rx^5 + {}^2/_9 x^4(0.8 + {}^1/_6) + {}^{11}/_{12} x^3 + {}^{16}/_9 Rx^2 + {}^7/_9 x^1 + {}^1/_6 Rx^0 + 4 = 59.88$$

$x = 2.4989$, not 2.501251, the solution for both the kaon and tau decays, but I will use 2.501251 below.

The chosen measured values are

$(0.16 + {}^1/_{70})x^5 = 14.27$	$K^{\pm}\pi^{\pm}\pi^0$	14.4 (.5)
$Rx^5 0.04 = 3.947$	$K^{\pm}\pi^{\pm}$	3.947 (.030)
${}^2/_9\, x^4(0.4 + 125^{-1}) = 3.549$	$K^{\pm}e^{\pm}$	3.549 (0.26)
${}^2/_9\, x^4(0.4 - 125^{-1}) = 3.41$	$K^{\pm}\mu^{\pm}$	3.41 (0.4)
${}^2/_9\, x^4\, {}^1/_6 = 1.45$	$K_L^0\pi^{\pm}e^{\pm}$	1.44 (0.4)
${}^1/_{18}\, x^3 + 0.1x^2 = 1.50$	$\pi^{\pm}\pi^{\pm}\pi^0$	1.49 (.06)
${}^{17}/_{36}\, x^3 + 0.2 \cdot 4 = 8.19$	$K^-2\pi^+\pi^-$	8.22 (.14)
${}^1/_{70}\, Rx^5 + {}^2/_9\, x^3 + R\, {}^x/_6 = 5.31$	$K_S^0\pi^{\pm}\pi^{\pm}\pi^0$	5.2 (.6)
${}^{11}/_{18}\, x^2 + {}^1/_6\, Rx^0 = 8.77$	$K^{\pm}\pi^{\pm}2\pi^0$	8.86 (.23)
${}^1/_6\, x^3 = 2.61$	$K_S^0\pi^{\pm}\pi^{\pm}$	2.80 (.18)
$Rx^2 0.2 = 1.241$	$K_S^0\pi^0$	1.240 (.022)
$0.1x^2 + {}^1/_6\, x^1 = 1.04$	$\pi^{\pm}\pi^{\pm}2\pi^0$	1.02 (.09)
$0.5x^1 + 0.75 \cdot 4 = 4.25$	$K^-2\pi^+\pi^-\pi^0$	4.3 (.4)

Here we see the incredible power of numerology. Only the coefficients can be adjusted in this model of neutral D mesons, and yet matching is almost perfect. There is some value in these equations, but without a clear theory, which also restricts the coefficients, we cannot depend upon the accuracy of the estimations.

12.3 Conclusion

The methods developed in chapter 11 on kaon branching fractions cannot easily be adapted to other particles, other than the tau, even if most of the major decays are known. Unlike the kaon probabilities, intermediate decays are sometimes counted separately. An additional complication is that two separate channels are often joined. Usually, nature splits this result in integer proportions. That is perhaps the only guide in developing the equations.

The estimation attempts for the D mesons show clearly that it is possible to obtain reasonable and very accurate results using only numerology or even numerology with some theoretical limitations. It also shows that even the most accurate results, when they cannot be supported by a consistent theory that does not depend on those results, should be vigorously questioned.

Chapter 13

Bohr's Atom

Bohr assumed that, in hydrogen, an electron orbits a proton. The orbits could only be at the specific distances that correspond to the energies found by Rydberg, who had generalized the Balmer sequence that had described visual light from hydrogen. A real electron cannot orbit the nucleus in an atom without emitting Larmor radiation. This problem was resolved by suggesting that the square of the wave function of the electron constitutes the probability of locating the electron in a specific volume of space. This solution created a new problem—namely, that the electron cannot be said to be in any location until measured. When we measure an individual electron, we always find a normal electron, but when an electron is not being measured, it is somehow spread throughout space. In a hydrogen atom, the electron is smeared out throughout the surface of a sphere whose distance from the nucleus of the hydrogen atom is associated with the electron's electrostatic potential energy.

There is something unusual about the equilibrium between the electron and the proton in the hydrogen rest state. The energy given off by an electron reaching that rest state is equal to the kinetic energy of an electron with a velocity of αc. The fine structure constant, α, plays an important role in physics. What can the rest state of hydrogen tell us about the mystery of α?

In SUiTe there is a Planck force that resists the Coulomb force and prevents the electron from reaching the proton in hydrogen. An electron, while moving toward a proton that it will join to form a hydrogen atom, compresses space in front of it. The proton serves as a backstop for the compressed duals of space. The duals are trapped between the proton and the electron and kept under compression by the electrostatic attraction of the electron for the proton. The Planck force will be the subject of section 13.4.

The rest state of hydrogen leads us toward the Bohr radius and the Bohr magneton. Both will be described in more detail in this chapter. They should also be able to help us better understand the fine structure constant. Let us imagine that the compression of space between the electron and the proton in hydrogen is spring-like, but because the electrostatic force is an inverse-square-law force, it is better to use a buoyancy model to describe the atom rather than the spring formula.

In buoyancy, gravity, which is also an inverse-square-law force, pulls an object into the water against a buoyancy force that is provided by the water. Space here will play the role of water, though the analogy with water buoyancy is limited. Space overall does not get significantly denser the closer the electron is to the proton, but the column of space trapped between the proton and the electron does get denser the closer the electron comes. The way in which a column of space becomes denser is not like an increase in the density of water, but we will model this increase in density as if it was a buoyancy force knowing that this model is only useful for the rest state of hydrogen. The equation for buoyancy is:

$$F_B = g\rho V_d$$

where g is the acceleration due to gravity, ρ is the density of the fluid that creates buoyancy, and V_d is the volume of the fluid displaced by the object that feels the buoyancy force.

We want to find an equilibrium between forces that will keep the electron connected to the proton in hydrogen, while at the same time preventing the electron from joining the proton in the nucleus. The simplest choice for the opposing force is $\hbar c/r^2$. The force that $\hbar c/r^2$ opposes is the electrostatic force, which is an inverse-square-law force similar to gravitation. I will introduce a true Planck force in section 13.4. The development of the buoyancy model only has value in showing that the electron in hydrogen need not be orbiting the proton. This explanation is limited to hydrogen.

The Planck constant was created to measure the energy of a photon of light. Unlike matter, light always travels at the speed of light. In the same way as I did with the special theory of relativity, a velocity squared will be considered as the pressure divided by the density of space. We now want to use the Planck constant to create a force using an electron, even though the electron is not moving at the speed of light as light does, so the use of the speed of light for the electron will not be appropriate. This will require us to insert a factor into our equation, v^2/c^2, the normalized pressure-divided-by-density ratio of space, which allows us to adjust the force opposing the electrostatic attraction between the electron and the nucleus. Our buoyancy equation is then:

$$k_C e^2/r^2 = (\hbar c/m_e r^2)\ (v^2/c^2)\ \rho_S V_e$$

The left-hand side is the electrostatic force of attraction between the electron and proton at a distance of *r*. The first term on the right-hand side is the Planck acceleration for the electron. It replaces the gravitational acceleration in the buoyancy equation. The second term is a necessary adjustment because the electron is not traveling at the speed of light. The last term is the density of space times the volume displacement of space by the electron. Using the rule we have from the special theory of relativity in SUiTe, we can rewrite v^2 as a pressure over a density, in this case P_B/ρ_S, where P_B is the buoyancy pressure. Now we have:

$$ke^2/r^2 = (\hbar c/m_e r^2)\ (1/c^2)\ P_B\ V_e$$

Solving for $P_B\ V_e$ and using the formula for the fine-structure constant $\alpha = ke^2/\hbar c$, we have:

$$P_B\ V_e = \alpha m_e c^2$$

or, in terms of α,

$$\alpha = P_B\ V_e/m_e c^2$$

The Bohr energy emitted by an electron as it is trapped by a proton and forms hydrogen can be stated in a form that looks very much like the kinetic energy that an electron would have at a velocity of αc:

$$E_{Bohr} = (1/2)\, m_e\, (\alpha c)^2$$

We can substitute αc for the velocity in our previous equilibrium equation for the buoyancy of the electron in hydrogen, and we will now have:

$$ke^2/r^2 = (\hbar c/m_e r^2)\, ((\alpha c)^2/c^2)\, \rho_S V_e$$

which leads to:

$$\alpha = \alpha^2\, (\rho_S\, V_e/m_e)$$

Solving for α, we have:

$$\alpha = m_e/\rho_S V_e$$

We now have two different values for α, one based on the buoyancy formula alone, the other incorporating Bohr's value for the velocity associated with the kinetic energy of the electron that is emitted as light when an electron and a proton come together to form hydrogen. In SUiTe, this energy would be the kinetic energy lost by the electron as bremsstrahlung (braking radiation) because the electron was forced to stop by the compression of space between it and the proton. We can now test to see whether these two definitions of α are consistent:

$$\alpha^2 = (m_e/\rho_S V_e)\, (P_B\, V_e/m_e c^2) = (P_B/\rho_S)(1/c^2) = v^2/c^2 = \alpha^2$$

The use of the buoyancy formula will result in a solution consistent with the Bohr solution for the rest state of the hydrogen atom, but the interpretation is different. Here the electron is not in an orbit or smeared out in a probability cloud, but it has a specific location relative to the proton that we cannot measure but that must nevertheless exist. The radiation emitted by a hydrogen atom is the bremsstrahlung of the electron as it decelerates while compressing space. The compression of space can be described using the Planck constant. The fine-structure constant is the ratio between the electrostatic force and another force that is relevant for two attracting electric charges forced apart by the compression of space between them. The use of the buoyancy formula, successful here at the equilibrium, provides us with one of the uses of the fine structure constant. We will find that the fine structure constant is like the elephant examined by the blind. We may have found the trunk, or the leg, or perhaps the tail, but we still don't understand the elephant.

13.1 Shells

In the Bohr model, the distance between the nucleus and the electron in hydrogen advances according to the square of the shell number times the Bohr radius. The Bohr radius is 52.9 picometers. The second shell is then 211.6 picometers. This is necessary because the Bohr energy levels associated with the shell numbers are one-half of the electrostatic potential energy at the distance of the shell. The corresponding Rydberg energies change according to the inverse-square of the distance from the nucleus. For the Bohr model to have the correct energies that match the Rydberg energies, it is necessary to square the distance in the Bohr model by squaring the shell number before multiplying by the Bohr radius.

In the buoyancy model, the energies associated with the equilibria, like the Rydberg energies, change according to the inverse-square of the distance. This allows the second shell to be twice as far as the first shell rather than four times as far, and the third shell will be three times as far rather than nine times.

The Bohr model is only valid for hydrogen and hydrogenic atoms (single-electron atomic ions). Why shouldn't the shell distances be valid for all elements, not just hydrogen? The shell distances for those atomic shells that are filled increase linearly and do not increase quadratically.

Looking primarily at the noble gases, the shell size is 28 pm. The covalent radii of the alkali metals are largest and are followed by a decline to the noble gas in that row of the table of the elements. The errors of the covalent radii are 3 pm or larger. This is about one eighth of the shell size. It is then possible to produce a table where the numbers should be multiplied by 3.5 pm and added to the shell distance for that row. For example, the covalent radius for Ca (second entry row three) is $4 \cdot 28 + 18 \cdot 3.5 = 175 \cong$

$176 \pm 10\,pm$. The table starts from the second shell with lithium and goes to the sixth shell that begins with cesium. After reaching 0 at tantalum, the radius decreases below the shell basic distance of 168. This last row that ends at radon is difficult.

20	12	8	6	4	3	2	0										
24	16	**11**	8	7	6	5	6										
26	18	16	14	12	**8**	8	6	4	4	**6**	3	3	3	2	2	2	0
24	16	14	10	7	4	2	2	0	0	**1**	0	0	0	0	0	0	0
22	14	**13**	10	10	10	9	9	9	8	8	7	7	6	6	5	**5**	2

One can see some clear patterns in the numbers. One can often find the value in the second or third row by doubling the number in the previous row and shifting left. There is a drop of 28 pm between the first two elements in a row. The *p* subshell becomes increasing flat until it is all 0s in the fifth row. The *f* subshell is nearly flat in the sixth row. The bolded numbers are the ones that seem out of place. *Al* and *Cr* have smaller radii than expected. *La* is larger. The increase in *Cu* and *Ag* match the subshell patterns $3d^{10}4s^1$ and $4d^{10}5s^1$ respectively. *Lu* is similar with a radius increase for the pattern $4f^{14}5d^1$.

Can the buoyancy equation, when applied to the hydrogen atom, have more than one equilibrium? In SUiTe, forces act along lines. If the electrostatic forceline is not pointed directly at the nucleus, the equation changes:

$$-ke^2 \cos\theta/(nr)^2 = (\hbar c/m_e(nr)^2)\,(1/c^2)\,P_B\,V_e$$

We still cannot determine from this equation how the proposed force, $\hbar c/r^2$, depends on distance, but it is possible that there are unstable equilibria at each node associated with a half-wavelength where the wavelength for hydrogen is equal to the Bohr radius. The electron will sit in the node because this is the location of lowest energy. The equilibria will be unstable because there is a lower energy state as the angle θ moves toward 180 degrees, and the electron moves closer to the nucleus. When θ is at 90 degrees, there is no attraction between the electron and the proton, and the electron will fly away, ionizing the hydrogen atom. This reflects the fact that, in SUiTe, the electrostatic force of the electron does not emanate in all directions but along a line. The solution to this problem is the

Planck force, a new force which keeps the electron out of the nucleus.

13.2 Subshells

We have found that the principal electron shells for hydrogen match, or are close to, the filled shells of other elements. In the case of hydrogen or helium at rest, the electrons rest on the principle shell closest to the nucleus. After helium, shells break down into subshells in the presence of a magnetic field. For our purposes, an electron can be considered in the shape of a ┼, where the long line is the electrostatic forceline and points toward the nucleus. The bar is made up of weak-force forcelines and is perpendicular to the nucleus and tangent to the shell surface.

First we need to consider how many electrons can fit in a shell. Since we know that the second shell is at two times the radius of the first, and since 2 electrons can fit in the first shell (helium), 8 electrons will be able to fit into the second shell because the second shell has four times the area of the first. The third shell has 9 times the area, hence 18 electrons. The fourth has 16 times the area and can fit 32 electrons.

Now we can ask what would happen if the atom was exposed to a magnetic field. The spins of electrons arrange themselves in a magnetic field, so that they are either parallel with the field or perpendicular to the field. With only two electrons, they can be opposite each other on the shell, as is the case with helium. We can arrange six electrons such that the bars indicating the spins of the ┼ electrons are all perpendicular to each other. Four will be along the equator and two at the poles. This gives a subshell holding 6 electrons. We can't add two more and still have each electron perpendicular to its neighbor. The last two of the 8 possible electrons in the second shell will form a separate subshell. The second shell will then have two subshells of 2 and 6 electrons.

To create the third subshell, we place four electrons above the equator and another four below, along with two electrons for the poles. Here we can see the simplicity and genius of nature. We are concerned with a single property of the electron, its weak-force forcelines. Nature must position neighboring electrons along a spherical surface such that their weak-force forcelines are perpendicular to each other. This is only possible for 6, 10, 14, 18, etc., consisting of 2 electrons at the poles plus an integer times 4 electrons. The number of subshells that can be in a shell are determined by which set of subshells add up to the maximum number of electrons that the shell can hold.

To better understand what nature is doing, take a tennis ball and a marker. Hold the tennis ball between your thumb and index finger. These will be the poles. Draw a short line at the equator that is either parallel or perpendicular to a line of longitude. Now draw lines the same size perpendicular to the first line about one-third of the way toward either pole. Draw two more lines that are perpendicular to the first line one-fourth of the way around the ball along the equator to both the left and right. Add lines above and below these new equator lines that are perpendicular to them. Draw in the last line along the equator on the opposite side of the tennis ball with the same orientation as the initial line. Draw in lines above and below perpendicular to that line. You will now have 12 electrons that are all perpendicular to their adjacent electrons. Finally, draw lines at the poles that are perpendicular to the four lines adjacent to the poles. This is the shell that holds 14 electrons, the *f* shell. Every electron is perpendicular to its 4 neighboring electrons.

After the first line had been drawn on the tennis ball, there was no longer any choice as to the positioning of the weak-force forcelines for the other electrons. It is this remarkable feature of electrons in atoms which is now being exploited in quantum computing. A single change to one electron results in cascading changes to all of the other electrons. These changes take almost no energy and happen at nearly the speed of light.

The weak-force forcelines of the electrons are better known as *spin*. However, the term "polarity" would have been a better choice because the electron is not spinning. Following common practice, I will use the term "spin," but please note that this use of "spin" has nothing to do with the real spinning of an electron that results in a magnetic field thread. Hopefully, the context will allow you to differentiate between these two uses of the word "spin."

F Electron Subshell

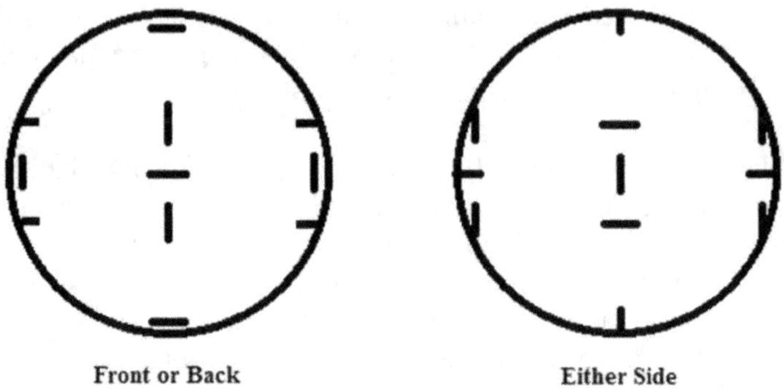

13.3 Bohr Magneton

If an external magnetic field is applied to a hydrogen atom, then the single energy level that is seen in the emission of a photon is now broken up into energies slightly larger or smaller than expected. This energy difference is equal to an integer times the Bohr magneton times the strength, in teslas, of the magnetic field. The integer ranges between 0 and the maximum number of shells jumped. The change in energy of a jump between adjacent shells can either add to, or subtract from, the total energy emitted when the electron stops. For example, an electron in hydrogen in the fourth shell—which would usually have only one possible energy for the emitted photon when it stops at the first shell—has seven possible energies in the presence of a magnetic field. Why are the energies quantized, and why doesn't it matter in what direction the magnetic field is pointing? We will need to answer both questions.

Electrons align in a magnetic field. Precisely how electrons align is discussed at length in chapter 14, on the Stern-Gerlach apparatus. What is interesting about electrons in atoms is that the set of spherical surfaces on which the electrons can be found—the shells—is limited. In SUiTe, the shells are equidistant for the different excited states of the hydrogen atom. The imposition of a magnetic field causes the hydrogen atom to rotate such that the combined electron-proton system is either perpendicular or parallel to the magnetic field. This process is discussed in the next chapter.

If the electron moves between shells, then the magnetic field will act on the electron. The more shells the electron crosses, the greater the distance over which the magnetic field can act and the greater the change in energy. We would expect that the energy depends on the strength of the magnetic field, and it does. This leaves us with only the Bohr magneton to explain, but before that, we need to understand why the change in energy of the emitted photon depends linearly on the number of shells the electron crosses. Normally, an electron passing through a magnetic field curves, and that would require a nonlinear explanation.

In SUiTe, forces act along lines. In the case of charged particles, the charged particles move in the direction in which their electrostatic forcelines are pointing. Although the magnetic fields in SUiTe cannot carry the electrostatic force, the spinning particles of space that make up the field carry the weak force that interacts with the electrostatic forceline of the electron, causing it to bend away from the direction in which it was moving. In SUiTe, like forces interact on the plane of their interaction, whereas unlike forces push or pull out of the plane of their interaction.

When a hydrogen atom is in its rest-state, its one electron has its electrostatic forceline pointing directly at the nucleus. The electron is somewhere on the first shell, one-half the Bohr radius from the nucleus. If the hydrogen atom is not in its rest-state, the electron is not on the first shell but is located on a higher-numbered shell that is some integer times one-half the Bohr radius away. When that happens, the Planck force that prevents the electron from entering the nucleus (section 13.4) and the electrostatic force are able to equilibrate if the electron points its electrostatic forceline at a slight angle away from the nucleus.

The electrostatic force drawing the electron toward the nucleus then becomes $-k_C e^2 \cos\theta / r^2$ where θ is the angle between the excited position of the forceline and the rest position, which points directly at the nucleus. Because the position of the electrostatic forceline of the electron can be anywhere along a circle that is in the plane perpendicular to a line joining the electron and the nucleus, θ can be said to describe a cone.

When a magnetic field is applied to a hydrogenic atom, as described in the previous section on subshells, the spin of the electron must be either perpendicular or parallel to the magnetic field. This results in the electrostatic forceline of the electron pointing in one of only four possible directions rather than toward anywhere along the edge of the base of the cone. For the purpose of determining the Bohr magneton, we will consider only two of those locations to which the electron is pointing, to the left and to the right on the circle the cone makes where these locations are perpendicular to the external magnetic field. This situation is illustrated below.

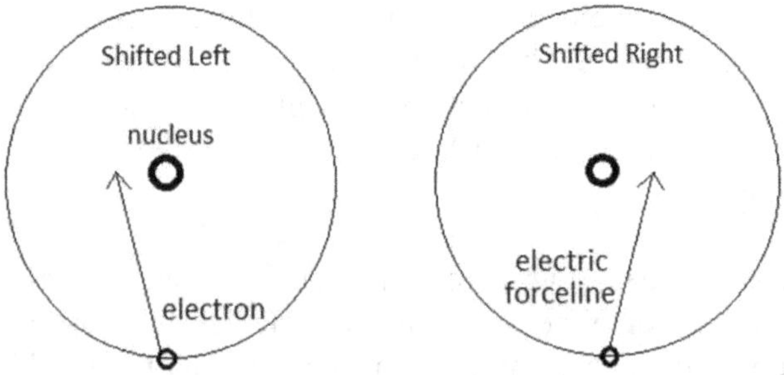

View of an excited hydrogen atom from the source of a magnetic field

Normally, the magnetic force would cause the electron to curl as it moves, if it is moving perpendicular to the field. Because the force of the magnetic field is perpendicular to both the field and the direction of motion, the curl will either help tilt the electrostatic forceline in the direction of the nucleus or away from the nucleus, depending on whether the electrostatic forceline was to the left or the right of the nucleus.

The magnetic force makes it easier for the electrostatic forceline to pull the electron toward the nucleus whenever the curl is in the direction of the nucleus. If the curl is away from the nucleus, the electrostatic force drawing the electron toward the nucleus will also have to overcome the magnetic force trying to pull it away from the nucleus. When the magnetic force joins with the electrostatic force, the electron accelerates more rapidly toward the nucleus, and more energy is released when it decelerates. If the magnetic force opposes the electrostatic force, the electron accelerates more slowly and less energy is released when it decelerates. When the electrostatic forceline is parallel to the magnetic field, the field has no effect. For deriving the Bohr magneton, we will explore the cases where the electrostatic forceline is perpendicular to the field.

There is a simple formula for the force of a magnetic field in terms of charge, velocity, and the field strength, if the magnetic field is perpendicular to the path of the charged object:

$$F = evB$$

where *F* is the force, *e* is a single electric charge, *v* is the velocity of the single charge, and *B* is the magnitude of the magnetic field. As was the case earlier, we are interested in the energy: force times distance. Distance is always an integer number of shells. We will work out the solution for the movement of an electron between adjacent shells. The shell-to-shell distance is r_h,, one-half of the Bohr radius. The Bohr radius is $\hbar/m_e\alpha c$, where m_e is the mass of the electron and α is the fine-structure constant.

$$E = F \cdot d = evBd = Bv(e\hbar/2m_e\alpha c) = B(e\hbar/2m_e)(v/\alpha c)$$

Experimentally, we know the answer for a one-shell shift by an electron that is part of a hydrogen atom in an external magnetic field, and it is $B(e\hbar/2m_e)$. This value, $(e\hbar/2m_e)$, is the Bohr magneton. Note that the 2 in the denominator for the Bohr magneton comes from using one-half the Bohr radius. From the equation above, we see that *v* must always be equal to αc

regardless of which shell-to-shell transition is involved, and even if the transition involves more than one shell. The velocity, which we know must change depending on how far the electron travels because the electron is accelerating, does not change at all in the equation. The velocity is always equal to its maximum, the velocity that the electron reaches when it comes from outside of the atom all the way to the first shell, the rest-state for hydrogen.

In SUiTe, space is discrete and magnetic fields are made up of threads, each one spinning at exactly the same rate. A stronger magnetic field has a higher density of threads. The magnetic force will then be the number of threads that the electron encounters. The faster it is moving, the more threads per second and the greater the force. Now consider a particle going a specific distance in a magnetic field with magnitude B. It doesn't matter how fast it is going because as long as it travels the same distance, it will encounter the same number of threads.

Here is where the curl of a charged object in a magnetic field becomes important. The faster the electron is moving, the more it will curl in the magnetic field, and the more magnetic threads it will encounter because its path in the field becomes longer. In the case of an electron moving in the atom in response to its attraction to the nucleus, the path does not change because the electrostatic force is much stronger than the external magnetic force at the distances encountered in the atom. This is why the velocity of the electron doesn't matter in our equation above. As long as it travels the same distance, it will encounter the same number of threads.

The electrostatic forceline can be either to the left or to the right of the nucleus from the point of view of the magnetic field, while the direction of the field determines whether the curl will be to the right or to the left. Therefore any magnetic field perpendicular to the path of the electron can create a curl force that can either speed up the electron or slow it down, depending on which side of the nucleus the electron's electrostatic forceline is pointing and which direction the magnetic field is pointing.

When the electron moves an integer number, n, of shells, we find experimentally that the previous energy of a photon in the absence of a magnetic field traveling n shells is split in the presence of a magnetic field into $2n + 1$ values. In the standard model, this is easily handled by imagining that the electron has specific quantum numbers associated with its orbit around the nucleus. SUiTe rejects the idea that the electron can orbit the nucleus, so the quantum numbers cannot be used to explain the energy splitting seen in the presence of a magnetic field in SUiTe. There is an alternate way of explaining Zeeman splitting without orbiting electrons creating a magnetic field, but it is cumbersome. I won't include it in this description of SUiTe.

If we choose to follow the explanation of Zeeman splitting in terms of quantum numbers, we have a magnetic quantum number, m, which ranges from $n - 1$ to $-n + 1$ where n is the shell number. For example, if $n = 3$, we can have $m = 2,1,0,-1,-2$. The possible changes in energy for emitted photons from a hydrogen atom due to an external magnetic field are then $\Delta E = m \cdot \mu_B B$, where $\mu_B = {}^{e\hbar}/_{2m}$ is the Bohr magneton and m is the magnetic quantum number.

Perhaps you can see why I was attracted to one-half of the Bohr radius. This radius is not only experimentally correct, but gives a simple and direct solution to the release of energy by an excited hydrogen atom in an external magnetic field.

13.4 The Planck Force

I will assume that the Planck force is an inverse-square-law force divided by the distance between w-duals squared. This value, when multiplied by the number of w-duals in a given volume of space between the electron and the proton in hydrogen, should equal the Coulomb force of electrostatic attraction between an electron and a proton, plus the background pressure that space already has. I will start with the volume of a cylinder with an area equal to the value of A_{pr} and height equal to r_h where r_h is one-half the Bohr radius and A_{pr} is the equatorial area of a protino (see section 20.5) and is also the area of a cylinder which can enclose an electron. Its value is taken from section 21.1. I will multiply the volume by the number of w-duals per cubic meter, w^3. The inverse of w^3 is r_w^3, the cube of the distance between w-duals. The Planck force is given by this equation:

$$F_P = \frac{h^2 c^2 \mu_w}{\pi}$$

μ_w has the numerical value of μ_0, the magnetic permeability. μ_w/π is the weak force permittivity of space in SUiTe, and the units are those of the electric permittivity of space, $s^2 e^2 kg^{-1} m^{-3}$, without the electric charge squared of e^2. After multiplication by the number of particles trapped between the electron and the proton in hydrogen, F_P will have the same units that k_C does after k_C is multiplied by two charges. The number of the w-duals of space compressed between the two attractive electric charges takes on the role of the missing squared electric charge that is necessary for the Coulomb force.

The equation that equilibrates the forces such that the electron does not go into the nucleus of hydrogen is:

$$\frac{F_P}{r_w^2} \cdot V_h \cdot w^3 = \frac{k_C e^2}{r_h^2} + \frac{F_P}{r_b^2} \cdot V_h \cdot w_b^3$$

The second term on the right-hand-side is the force due to the background pressure of space which must be added to the additional force resulting from the electrostatic force.

$$F_P \cdot V_h \cdot \left(w^5 - w_b^5\right) = \frac{k_C e^2}{r_h^2}$$

$$\left(w^5 - w_b^5\right) = \frac{k_C e^2}{r_h^3 F_P A_{pr}} = \frac{k_C e^2 8 m_e^3 \alpha^3 c^3 \pi}{\hbar^3 h^2 c^2 \mu_w A_{pr}} = \frac{2\alpha^4 m_e^3 c^2}{\pi \hbar^4 \mu_w A_{pr}}$$

The distance between w-duals in uncompressed and uncontracted space from the equation linking the constants in section 21.1 is about $8.60257 \cdot 10^{-19}$ meters or 1.16244 w-duals per attometer. The distance between w-duals for the rest-state of hydrogen in the column of space between the electron and the nucleus, r_w, is then $7.294 \cdot 10^{-19}$ meters or 1.371 w-duals per attometer.

The background value for the distance between w-duals must be greater than the compressed value in the hydrogen atom, but it cannot be much more because in an extreme hydrogenic atom, such as lead with a single electron, the space within the column between the electron and the nucleus would become too compressed. Such a compression in space would be measurable.

For Z equal to the number of protons in a hydrogenic atom, the distance between the electron and the proton is reduced to $Z^{-1}r_h$. The cylinder of compressed w-duals between the electron and the nucleus retains the same area, which depends only on the electron. The volume is smaller, but the number of w-duals increases in spite of having less space, and the force of repulsion between the w-duals that are compressed increases. The resulting equations for determining w_Z are then

$$\frac{F_P}{r_{w,Z}^2} \cdot \frac{V_h}{Z} \cdot w_Z^3 = \frac{Z^2 \cdot k_C Z e^2}{r_h^2} + \frac{F_P}{r_b^2} \cdot \frac{V_h}{Z} \cdot w_b^3$$

$$\left(w_Z^5 - w_b^5\right) = \frac{Z^4 k_C e^2}{r_h^3 F_P A_{pr}} = \frac{2Z^4 \alpha^4 m_e^3 c^2}{\pi \hbar^4 \mu_w A_{pr}}$$

The following is a table of the distances between the w-duals of space and the number of w-duals per meter. The more protons that are in the hydrogenic atom, the greater the Coulomb force will be.

		$r_{w,Z}$	w_Z
H	Z=1	$7.294 \cdot 10^{-19}$	$1.371 \cdot 10^{18}$
He	Z=2	$4.657 \cdot 10^{-19}$	$2.147 \cdot 10^{18}$
Li	Z=3	$3.393 \cdot 10^{-19}$	$2.948 \cdot 10^{18}$
Be	Z=4	$2.699 \cdot 10^{-19}$	$3.706 \cdot 10^{18}$
B	Z=5	$2.258 \cdot 10^{-19}$	$4.428 \cdot 10^{18}$
C	Z=6	$1.952 \cdot 10^{-19}$	$5.123 \cdot 10^{18}$
N	Z=7	$1.726 \cdot 10^{-19}$	$5.795 \cdot 10^{18}$

O	Z=8	$1.551 \cdot 10^{-19}$	$6.448 \cdot 10^{18}$
Ca	Z=20	$7.451 \cdot 10^{-20}$	$1.342 \cdot 10^{19}$
Pb	Z=82	$2.410 \cdot 10^{-20}$	$4.150 \cdot 10^{19}$

The extraordinary thing in the above table is that a more than 550,000-fold increase in the electrostatic force between the single electron and the nucleus of hydrogenic lead resulted in only a 30-fold decrease in the distance between w-duals. Space is already compressed, and very small changes in the compression of space result in large changes in the ability of space to resist further compression.

There are also semi-stable equilibria for the hydrogen atom at the shells associated with various excited states. In the Bohr model, the equilibria are at squared multiples of the Bohr radius, 1, 4, 9, 16, 25, ... n^2, where n is the shell number. In SUiTe, the equilibria are at integer multiples of one half of the Bohr radius. For the electrostatic force to be stable at those distances, rather than the square of an integer times a radius, it is necessary for the electrostatic force to act along a line that intersects at an angle with electrostatic lines from the proton. The angle is between the tilted electron forceline and a direct line from the electron to the proton. The cosine of the angle equals $1/n$. The electrostatic force in hydrogen is then divided by n. Including the change in distance, which has been multiplied by n, the equations become

$$E_n = \frac{k_C e^2}{4n^2 r_h} \qquad F_n = \frac{k_C e^2}{n^3 r_h^2}$$

where E_n is the energy of the emitted photon as bremsstrahlung from an electron arriving from infinity to the n shell.

Like the Coulomb force, the Planck force is also reduced because of the tilting of the electron away from the nucleus. The lines of w-duals in the column that stretches from the electron to the proton are now striking the electron at an angle rather than dead on. Dividing by n will adjust for a single line, but to adjust for the surface formed by the end of the column that is pushing against the electron, it is necessary to divide the Planck force by

n^2 rather than n as is the case with the Coulomb constant. The Planck force is then:

$$F_{P,n} = \frac{h^2 c^2 \mu_w}{n^2 \pi}$$

The subsequent forces are:

$$\frac{F_{P,n}}{r_{w,n}^2} \cdot nV_h \cdot w_n^3 = \frac{k_C e^2}{n^3 r_h^2} + \frac{F_{P,n}}{r_b^2} \cdot nV_h \cdot w_b^3$$

The resulting equation for determining the number of w-duals per attometer in excited hydrogen where n is the shell number is:

$$\left(w_n^5 - w_b^5\right) = \frac{k_C e^2}{n^4 r_h^3 F_{P,n} A_{pr}} = \frac{2\alpha^4 m_e^3 c^2}{n^2 \pi \hbar^4 \mu_w A_{pr}}$$

After adding an additional digit to the estimates in order to show that there remains a small difference between n=110 and the background, the resulting values for the distance between the w-duals in meters and the number of w-duals in a meter for the excited states of hydrogen are:

	$r_{w,n}$	w_n
n=1	$7.2944 \cdot 10^{-19}$	$1.3710 \cdot 10^{18}$
n=2	$8.1373 \cdot 10^{-19}$	$1.2289 \cdot 10^{18}$
n=3	$8.3764 \cdot 10^{-19}$	$1.1938 \cdot 10^{18}$
n=4	$8.4710 \cdot 10^{-19}$	$1.1805 \cdot 10^{18}$
n=5	$8.5169 \cdot 10^{-19}$	$1.1741 \cdot 10^{18}$
n=10	$8.5807 \cdot 10^{-19}$	$1.1654 \cdot 10^{18}$
n=110	$8.60238 \cdot 10^{-19}$	$1.16247 \cdot 10^{18}$
background	$8.60259 \cdot 10^{-19}$	$1.16244 \cdot 10^{18}$

$n = 110$ was chosen because the transition between 110 and 109, named H109α, results in a photon emission that is important for radio astronomy. As can be seen, there is not much difference between the background pressure of space and the compression of space between an electron in the 110th shell and the nucleus in hydrogen.

The calculation of the energy of an emitted photon is different in SUiTe. In the Bohr model, half of the potential energy is emitted and half goes to the kinetic energy of an orbiting electron. In SUiTe, one fourth the energy is emitted and three-fourths goes into the compression of space between the electron and the proton.

The use of half the Bohr's radius combined with the calculation of the radius of an excited hydrogen atom as n times half the Bohr radius rather than n^2 times the Bohr radius creates an interesting problem for the subsequent Coulomb force. In the Bohr model, the Coulomb force would be divided by n^4, but in SUiTe, it is divided by n^3. The Coulomb force of attraction within the hydrogen atom is stronger in SUiTe than in the Bohr model, but the energies of the emitted photons are the same.

Inverse Square Model or Spring Constant?

For a spring at low tension, a linear change of one unit distance requires an energy of one unit even as the spring continues to be compressed. Eventually, the spring will need much more than one unit of energy, and the energy needed to compress the spring will continue to increase rapidly. Taking the unit of distance to be one half the Bohr radius, the compression of space between the electron and the nucleus takes nearly constant energy from 99 radii to 69 radii. Dividing the energy necessary to compress space from 101 radii to 100 radii in comparison to 100 radii to 99 radii, the ratio is 1.020, or two percent greater. From 71—70—69, the ratio is 1.029. Even from 69 to 39, it goes from 1.029 to 1.051. After that, the compression of space is best explained by the Planck force as described earlier in this section. The equation for the change in energy needed by the electron in hydrogen to compress space by a step of $r_B/2$ is

$$E_{n+1,n} = k_C e \div \frac{r_B}{2} \cdot \left(\frac{1}{n(n+1)}\right) \cdot \left(1 - 0.25\left(\frac{1}{n} + \frac{1}{n+1}\right)\right)$$

where k_C is the Coulomb constant, e the elementary charge, n the shell number, and E is the energy in electron volts.

Chapter 14

Stern-Gerlach Apparatus

Two permanent magnets are used to create an inhomogeneous magnetic field between them. The lower one is flat and the upper one is wedge-shaped, with the sharp end of the wedge pointed toward the lower magnet. If we imagine the field in terms of threads, the threads reach up from the lower magnet toward the point of the upper one to form a triangular field with the density of threads lower at the base and highest at the point. The distance between the threads decreases toward the pointed magnet. This creates an inhomogeneous magnetic field that is stronger near the pointed magnet and weaker near the flat one. An additional electric field is applied in such a way that the usual curl of the electron in a magnetic field is prevented.

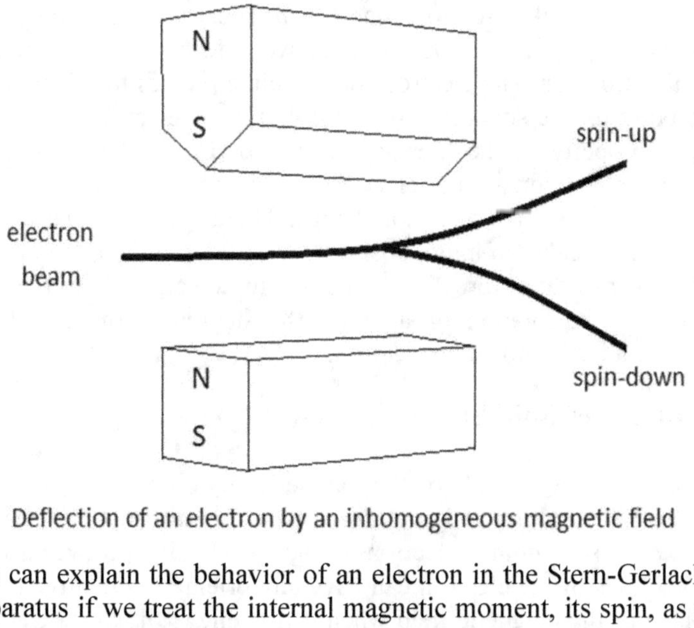

Deflection of an electron by an inhomogeneous magnetic field

We can explain the behavior of an electron in the Stern-Gerlach apparatus if we treat the internal magnetic moment, its spin, as a question of rotation in the plane perpendicular to its motion. Imagine the electron as two lines, with one pointing in the direction of motion and the other line perpendicular to motion. The first line provides the axis of rotation. The second line is the electron's spin. When electrons are really spinning, they create a thread of a magnetic field. The spin described here is the *polarity* of the electron, which is similar to the polarity of plane-polarized light.

The axis of rotation is the electrostatic forceline. Charged particles move in the direction in which their electrostatic forcelines are pointing. The line that is perpendicular to the electrostatic forceline is in fact composed of three weak-force forcelines which are approximately one fourth of the length of the electron apart from each other. In the electron, these weak-force forcelines are always perpendicular to the motion of the electron. Two are weak-force negative, and the middle one is weak-force positive. The w-duals of space are weak-force negative and are attracted to the one weak-force positive forceline of the electron, but repelled by the two weak-force negative forcelines.

The three weak-force forcelines of the electron can be imagined in the shape of a squat Z where the weak-force positive forceline is the crossbar. The electrostatic forceline pierces the Z from top to bottom. The squat Z is responsible for the electron's quantum spin property in the Stern-Gerlach experiment. This is not an intrinsic property like electrostatic charge. If we rotate an electron, its charge does not change. There is no possible way to change an electron into a positron. A 180 degree rotation around the electrostatic forceline turns an up electron into a down electron. A rotation of another 180 degrees turns the down electron back into an up electron.

14.1 Three Spin States but One Is Unstable

The magnetic threads of the magnetic field are made up of w-duals, all spinning and pointing in the same direction, ultimately linked to spinning electrons. The plane of the weak-force forcelines of the electron can have any orientation relative to the plane of the magnetic field when it first enters the Stern-Gerlach apparatus. The magnetic field acts upon the weak-force forcelines, so that they are either parallel or perpendicular to the magnetic field, generating a rotation of the electron around the electrostatic forceline.

There are three equilibrium states for the electron: spin-up, spin-down—both of which are parallel to the threads of the magnetic field—and one state that is perpendicular to the magnetic threads. This last state is unstable in a magnetic field. The two stable equilibria do not depend on whether the upper wedge-shaped magnet is a north pole or a south pole. These two parallel states are associated with a squat Z and a reversed squat Z. The squat Z, because of where the attractive weak-force positive forceline is pointing, pulls the electron up. The weak-force positive forceline in the reversed squat Z also pulls the electron up, but because it is at the back of the electron, the direction of the electron is deflected down. Since it is exactly the same force, with the only difference being that one force acts toward the front of the electron and the other at the back, the deflection is equal in both directions.

If we now keep only those electrons in which the crossbar in the squat Z is pointing up (the ones that were deflected up), and throw away the electrons that were deflected down, then we will have a stream of electrons whose spin orientation is known. We can pass that electron stream into a second Stern-Gerlach apparatus. If the second apparatus has the same orientation as the first, then we can guess that all the electrons will remain spin-up. If the second apparatus is perpendicular to the first, then all the electrons will need to rotate 90 degrees to become parallel to the magnetic threads of the second Stern-Gerlach apparatus. We can guess that half of the spin-up electrons will become spin-down because there is no preferred direction of rotation for the electron. Finally, we can guess that if the second apparatus is rotated 180 degrees with respect to the first, so that now the wedge-shaped magnet is the lower one and the flat magnet the upper one, then all the spin-up electrons will be deflected down.

The following diagram shows the entering electron as if it were a plane with the weak-force forcelines acting as wings, and how it will respond to the magnetic-field threads that it encounters if it enters the second Stern-Gerlach apparatus with its weak-force forcelines at 30 degrees from the plane of the magnetic field. I

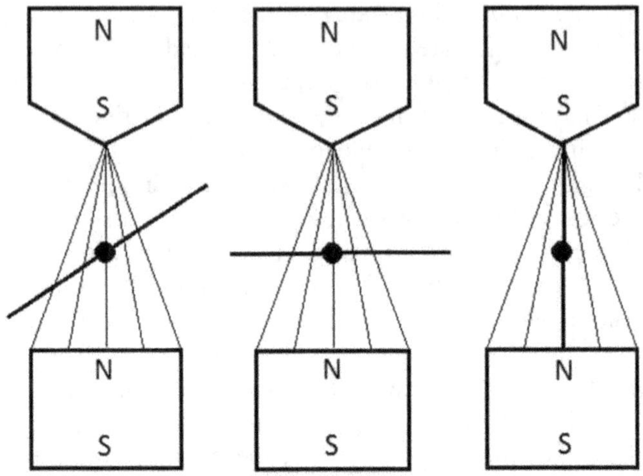

An electron enters at 30 degrees | The electron is pushed back to 90 degrees | The electron comes to rest at 180 degrees

show the case where the electron is initially rotated back to 90 degrees before then rotating another 90 degrees to become parallel to the magnetic field. It is also possible for the electron to rotate to a parallel state without first rotating back to 90 degrees.

14.2 Spin Probability with Two Apparatuses

What is the likelihood that a spin-up electron will remain a spin-up electron if the second apparatus is set to 30 degrees from the first? The classical expectation is that 100 percent of the spin-up electrons will remain spin-up for any setting of the second apparatus relative to the first between 270 and 90 degrees. The SUiTe expectation is different. There are three weak-force forcelines in SUiTe: two that are weak-force negative that will be repelled by the magnetic threads, and one that is weak-force positive and will be attracted to the magnetic threads. The weak-force positive forceline is slanted and will dominate the overall behavior of the electron, which is why there are two stable equilibria, both parallel to the magnetic field. The two weak-force negative forcelines cause a rotation toward the perpendicular unstable equilibrium.

As is always the case in SUiTe, the force of attraction is determined by the cosine of the angle. In this case the force of attraction will be the angle between the weak-force positive forceline as the electron leaves the first apparatus and the magnetic threads of the second apparatus—namely, 30 degrees. The cosine of 30 degrees is $\sqrt{3}/2$. This is the likelihood that a spin-up electron from the first apparatus will be counted as a spin-up electron in the second. The likelihood that it will not align is $1 - \sqrt{3}/2$. This is where things get interesting. If the electron is bumped away from the nearest equilibrium state to the point where the squat Z of weak-force forcelines is perpendicular to the magnetic threads, then there is still a 50 percent chance that it will be deflected back and be counted as spin-up. We now have:

$$\sqrt{3}/2 + 0.5(1 - \sqrt{3}/2) = 0.5 + (\sqrt{3}/2)/2 = 0.5 + \cos\theta/2$$

where θ is 30 degrees. The formula does not change for angles over 90 degrees. It can be rewritten as $(\cos\theta/2)^2$. This is the same formula that is used in quantum theory. For our example of 30 degrees, $(\cos\theta/2)^2 \cong .933$. With the Stern-Gerlach experiment, there is no need to treat the electron in a way that is inconsistent with classical expectations if we allow the electron to have a specific structure and allow the magnetic threads to be discrete and not distributed uniformly in space.

Why aren't electrons always being repelled or attracted to the w-duals of space, since, according to SUiTe, space is filled with spinning w-duals? The answer is that only the w-duals that are joined to a spinning electron in a magnet have enough force to be able to have an effect on a free electron passing through a magnetic field. One electron is in contact with the other through the link created by the w-duals. Type 1 superconductors can take advantage of these links between the electrons in the magnetic field and those in the superconductor in order to levitate objects.

Free electrons will always align their spin parallel to the magnetic field if their motion is perpendicular to the field, as was the case with the Stern-Gerlach experiment. In the Cooper pairs of a type 1 superconductor, the two electrons are locked together with their spin lines perpendicular to each other and are unable to rotate their spins into alignment with the field. This places pressure on the electrons in the magnet that are generating the field, pushing them away from the superconductor. One electron in the magnet is trying break up the two electrons that form the Cooper pair in the superconductor.

Chapter 15

De Broglie, Heisenberg, Schroedinger

According to de Broglie, particles behave like waves. According to Einstein, in his description of the photoelectric effect, light behaves like particles. In 1923, Compton showed that a photon could scatter an electron in the same way that a particle would, conserving momentum. The energy of the photon is *hf*, where *f* is the frequency of the photon and h is Planck's constant. We can then plug the energy into Einstein's equation $E = mc^2$ to get the apparent mass of the photon. An electron struck by a high-energy photon behaves as if it was struck by a particle of mass equal to hf/c^2. There is another way of writing Einstein's famous equation that is much better for light—namely, the energy-momentum relation $E^2 = p^2c^2 + m^2c^4$, where p is the relativistic momentum and m is the rest mass of a particle. Although m is the rest mass, with light there isn't any rest mass, so we have a very simple equation for the momentum of light, $E/c = p$. This is what Compton was able to prove.

15.1 Particles Are Waves

From the work of Compton and Planck, the energy of a photon can be described either in terms of its apparent mass and momentum, $E = mc^2 = pc$, or in terms of its frequency, $E = hf$. What if this is also true for matter when we think of matter as a wave? If we imagine that this is true for the electron, then its energy in terms of its wavelength would be $E = hf$, just as for light. If we were to accept this equation as true, then the momentum of matter, if it was moving at the speed of light, could be described in terms of its wavelength as $p = mc = h/\lambda$, where λ is the wavelength and m is the relativistic mass.

Matter cannot move at the speed of light, but to make matter as much like light as possible, de Broglie realized the relation $p = mv = h/\lambda$ would have to hold. No other mathematical relation could work. For light, the apparent mass is useful in calculations which emphasize the particle nature of light, such as the scattering of electrons by energetic photons. For matter, the de Broglie relation is used to find the wavelength for those calculations which emphasize the wave nature of matter.

Bohr had worked out the momentum necessary for an electron to avoid dropping into the nucleus of hydrogen for each possible orbit the electron could take. De Broglie could now show that the associated wavelength of the orbiting electron was the square of an integer multiplied by the Bohr circumference—2π times the Bohr radius. In the hydrogen atom, it was no longer necessary to think of the electron as orbiting and emitting Larmor radiation. Instead, the electron could now be thought of as a wave that somehow surrounded the proton, and whose wavelength was equal to the circumference of its orbit.

In SUiTe, the wave nature of the electron is provided by the weak-force forcelines that are perpendicular to the electrostatic forceline and, as a result, are perpendicular to the direction of motion of the electron. If I attempt to pass an electron through a slit in a barrier, then the weak-force forcelines of the moving electron will interact with the weak-force forcelines of the electrons in the atoms of the barrier. The ability of the edges of the slit to deflect the electron will depend on how close the electron is to the edge of the slit and on its momentum. The force exerted by the slit on the electron is independent of the speed of the electron, but the force required to deflect the electron will depend on the degree to which space is compressed in front of the electron—namely, its momentum. The faster the electron, the more compressed space is, and the greater the force that will be needed to change the direction of the electron. The wavelength of matter is inversely proportional to the momentum.

The wavelength of the electron is dependent on the momentum of the electron in the way described by de Broglie, but SUiTe varies in how that wavelength can be used. It is necessary for deflection, interference, and scattering, but when the electron is not moving, as is the case for electrons in atoms, the wavelength can no longer be used. The wavelength of matter in SUiTe is strictly associated with momentum, and, even then, only when one object is being affected by another. The wavelength of a free electron moving in space, for example, is of no use. In that sense, we should not consider the wavelength of matter as a property of matter, but as a necessary tool to describe how the interaction of the weak-force forcelines between objects affects their motion.

15.2 Nature Cannot Be Measured Exactly

The equation for the quantum uncertainty in a combined measurement of momentum and location is $\Delta p \Delta x \geq h$. In order to locate the particle, it is necessary to bounce a photon of light off of it. The wavelength of the photon of light has to be at least as small as the particle we wish to see; otherwise our error will be at least as large as the wavelength. The smaller the wavelength of light we are using, the more powerful the energy of the light, and the more accurately the location of the particle can be determined. But the exchange of energy between the photon and the particle we wish to measure causes the particle to recoil, modifying its momentum. The more powerful the light, the greater the recoil, and the more momentum that is transferred from the photon to the particle whose location and momentum we wish to measure. This particular explanation of Heisenberg uncertainty depends on disturbing the particle we wish to measure as we attempt to measure it.

There is another way to reach the same conclusion, assuming that we can somehow measure the particle without disturbing it. With a perfect microscope that can use anything we wish as a source for illuminating the particle, our only concern is the diffraction limit. The diffraction limit depends on the wavelength of the illumination and on the angle of the cone from the detector to the microscope lens and from there to the particle. The error in the measurement due to diffraction where q is the location of the particle is: $\Delta q \sim \lambda / 2\sin\theta$ where λ is the wavelength and θ is the angle of the cone.

There is also an error in measuring the momentum. We know that a photon entered the microscope, bounced off of the particle we wish to measure, and then went through the lens, but we don't know precisely at what position it went through the lens, which leads to a potential error of $2\sin\theta$. This gives the error

$\Delta p \sim 2 \sin\theta \, {}^{h}/_{\lambda}$ in our measurement of the momentum of the particle. The above relation for the momentum of the particle assumes that the magnitude of the particle's momentum is unchanged by its interaction with the photon. It is not necessary to assume that the direction of the particle has changed, either. It is only necessary to assume that the movement of the particle being struck by the photon generates an uncertainty in where the photon enters the lens after the collision.

Combining the two uncertainties gives: $\Delta p \Delta q \sim h$. Every assumption we have made was made in order to reduce any disturbance to the position, or the momentum of the particle, to 0. We note that the product of the errors is proportional to h. Because we have tried to minimize the error, any constant of proportionality must be greater than 1 in order to take into account the disturbance caused by the measurement or by a microscope lens that is less than perfect. This is why the uncertainty principle is usually written as $\Delta p \Delta q \geq h$.

This form of Heisenberg's uncertainty principle is unchanged in SUiTe. It represents the rules of classical physics applied to very small particles, where locating either the particle, or a photon that bounced off of the particle, depends on the diffraction limit of the lens, and the location where a photon entered the lens, because of the uncertainty associated with the momentum of the particle. Unfortunately, the uncertainty principle has become much more than a relation between measurement errors of location and momentum. There is a similar uncertainty relationship regarding time and energy: $\Delta E \Delta t \geq h$ which allows energy to be borrowed from space for very short periods of time. This isn't possible in SUiTe.

Momentum and position are real entities. Space is real, and it is possible to specify a position in space. The momentum of a particle can be determined by the slant of its forcelines. Momentum is not dependent on time. Nothing in our universe is. Time is either off or on. If it is on, there is a multiverse that is filled with myriad universes, and these universes are moving from birth toward death but not in terms of well-defined seconds. Clock time is what we use to predict the future. Nature lives in the never-ending now. The failure of the energy-time uncertainty principle in SUiTe prevents virtual particles, and also makes the bosons of the weak force impossible because of their huge mass.

15.3 Electrons Are Plane Waves

Given the frenzy of the 1920s in both physics and popular culture, it is surprising that over a year passed between de Broglie's new understanding of matter and Schroedinger's wave equation. This may have been because Schroedinger was searching for a relativistic wave equation—something that escaped him—but the equation he found is used frequently because when the particles involved are not relativistic, it provides acceptable answers. Like Heisenberg uncertainty, the equation can be described in terms of classical physics.

The wave function of a plane wave is: $\psi = \exp(i(kx - wt))$. The value of the wave function for some x and t is the amplitude at that point and time for a wave moving in direction x, where x is the displacement from some initial point x_0 and t is the elapsed time from an initial time t_0. The phase angle is $(kx - wt)$. In the case of a normal wave like water, but not a wave of light, the amplitude would give us the height of a crest or the depth of a trough. The phase angle would tell us what we could expect to see if two waves moving in the same direction, but with slightly different frequencies, moved through each other. If two water waves pass through each other in this manner, there will be a point at which their phase angles are 180 degrees out of phase. If both waves have the same maximum amplitude, the water will be completely flat at that point because the crest of one wave cancels the trough of the other.

If I assume that electrons are plane waves, I can describe them in terms of complex numbers, where the complex axis is perpendicular to the real axis. By using complex numbers, I can reduce the two dimensions of a plane wave of light into one complex dimension. A plane wave is $\psi = \exp(ikx - iwt)$ where ψ is the wave function, $k = {}^{2\pi}/_{\lambda}$ and $w = {}^{2\pi c}/_{\lambda}$. The central equation of classical physics is the Hamiltonian:

$$E = {p^2}/{2m} + V$$

The total energy (E) is equal to the kinetic energy $\left({p^2}/{2m}\right)$ plus the potential energy (V). I will assume that the wave is moving in the x direction.

Using derivatives of the wave function, I can now recreate the Hamiltonian with the assumption that electrons are plane waves.

$$i\hbar\frac{\partial}{\partial t}\psi(x,t) = -{\hbar^2}/{2m}\left(\frac{\partial^2}{\partial x^2}\right)\psi(x,t) + V(x,t)\psi(x,t)$$

This is the Schroedinger equation. An electron is wave-like, but it is not a true wave. Max Born noted that the square of the amplitude (the wave function) in the Schroedinger equation gives the probability of finding an electron in a small sphere centered at x. Unlike waves, electrons cannot combine troughs and crests and disappear. On the other hand, electrons can appear in a place that would be impossible under the classical assumptions for an electron. This can occur when the wave function has a non-zero value for a location on the other side of an electrostatic barrier, and is called *quantum tunneling*.

The photon is not a true wave either but is a pulse, and is particle-like. The square of the amplitude for the photon can be treated in the same way as for the electron, and the resulting theory agrees with experiment. The diffraction of a photon at the edge of an object can be described in exactly the same way as for an electron, and results from the same process. The weak forces at the slit retard one side of the weak-force forcelines of the electron more than the other side if the electron does not pass through the slit at the exact center. In the case of a photon, the w-duals that make up the photon are bent by the weak-force forcelines of the electrons that are part of the material at the edges of the slit. Like the electron, the bending of the bundle of weak-force forcelines that constitute a pulse of light changes the photon's direction. Because the force exerted by the slit does not change with the speed of the particle passing through the slit, the greater the momentum of the particle or photon, the less it is diffracted.

In the case of interference, in order to treat photons and electrons in the same way, the empty bands that result from passing electrons or photons through two slits in a barrier are locations where the probability of locating an electron or a photon is 0. Neither electrons nor photons cancel each other in the way waves can. Schroedinger's equation, though derived from classical physics, suggests a very different theory for electrons or photons than classical wave theory. This difference will be the subject of the following chapter.

Chapter 16

Two Slits and Bell's Inequalities

Experiments where light was passed through two narrow, closely-spaced slits were performed by Young in the early 1800s. Passing light through the slits created a series of light and dark bands. In order to create these bands, the light going through one slit must be able to modify the behavior of the light going through the other slit. This is *wave interference*. Newton's theory that light was made of particles could not explain this behavior. The wave theory of Huygens could.

Planck's explanation of blackbody radiation suggested that we could think of light as packets of energy. Einstein's explanation of the photoelectric effect, and Compton's explanation for the scattering of electrons by photons, both supported Planck. Packets of energy are closer to particles than to waves. A packet is not a wave-train stretching out 2 meters or more, as should be the case for visible light. The solution to the problem was a suggestion by Born; the wave function in Schroedinger's equation was the square root of the probability of locating a particle, the probability amplitude. The dark interference bands of the two-slit experiment were locations where electrons or photons had a low probability of being found.

Born's solution placed particles of light and matter on the same footing, making it possible to treat the interaction of light with matter in terms of probability. Heisenberg's uncertainty makes it clear that we can never measure the exact location and momentum of any particle simultaneously. There are still two interpretations of the Born solution for the wave function of the Schroedinger equation: (1) choosing the probabilistic approach because of our inability to observe what is truly happening to particles or light in the same way that Brownian motion appears to be probabilistic but is caused by unseen interactions with molecules, or (2) the idea that probability is a part of nature and there is no underlying layer that our theories are approximating.

SUiTe chooses (1), Einstein's view—namely, that there is an underlying layer. If we could somehow observe this underlying layer, we would not find inherent probability. Each new state is completely determined by the previous state. They represent the causes of the new state. Nature is deterministic.

Bell's inequalities are the direct consequence of the quantum theoretical view that two particles with a combined quantum state can nevertheless be separated in space. This combined quantum state is *entanglement*, and the consequence of entanglement is that modifying one of the entangled particles instantaneously modifies the other. In classical physics, all interactions are local. Any forces acting between particles cannot act faster than the speed of light.

The two-slit problem and Bell's inequalities are linked when we consider what happens when we send photons through the two slits one photon at a time. If we close one slit, the photon behaves according to the rules of diffraction. There are no bands of light and dark. When the second slit is reopened, the bands reappear even if the photons are being sent to the slits one at a time. The photon appears to interfere with itself, as if it had managed to go through both slits at once. With Bell's inequalities, the two entangled particles represent a single particle even though, classically, a single particle can't be split into two identical pieces an arbitrary distance apart.

The description SUiTe has for the two-slit experiment is possible, but not necessarily correct. It matches the experiments, but changes our view of all spontaneous decays, including the emission of light by an excited atom. SUiTe replaces a single photon that goes through both slits at the same time with two photons of the same frequency passing through the two slits where one is a sub-threshold photon and cannot be detected.

The description that SUiTe provides for Bell's inequalities depends on a remaining loophole in the tests of the inequalities that is associated with time. The resulting explanation is testable and will show, or not show, that SUiTe is correct about quantum entanglement.

16.1 Explanation of the Two-Slit Experiment

In the standard model, it is possible to think of all interactions as being between particles. The wave function becomes the probability amplitude which, when squared, is the probability of locating the particle. For the two-slit experiment, Feynman suggested, in a thought experiment, that a single photon could break itself into multiple copies, and these copies could take any path possible to the detector concurrently. Physically, a photon cannot become many photons, nor can it take multiple paths to the same location without considering the change in the distance to the detector associated with the path.

With the two-slit experiment, Feynman suggested that we consider the path length in terms of how it would change the phase of the photon, and then add up the individual phases of the imaginary photons traveling the different paths. In this approach, the length of time required by a photon to travel the path is not factored in. The phase of a photon can be modeled as a rotation in two dimensions, which can be reduced to one dimension using complex numbers. Using complex mathematics, we can add up the various paths. Most of the possible paths cancel each other. The ones remaining provide us with the probability that we will see a photon at the detector at a particular location. This process correctly describes the dark and light bands that we see. In classical physics, the phase corresponds to the changing amplitude of a wave, but in the standard model, the phase helps to determine the probability.

This wouldn't be possible in SUiTe because, in SUiTe, all particles are real. Photons cannot split into multiple photons, and the path the photon takes must be a real path where the elapsed time associated with the path matters. The standard model solution to the two-slit experiment requires that the probability that a photon will be detected in a band does not depend on the number of photons per second passing through the slits. Experimentally, the percentage of photons outside of the light bands increases with the time between photons. The separation between the light and dark bands becomes increasingly fuzzy, but the light bands remain visible.

16.2 The Two-Slit Experiment in SUiTe

Einstein may have been one of the first to complain about inherent probability in physics. Certainly, we use statistical mechanics to explain Boltzmann's work on the temperature of a gas. Einstein had carefully studied the work of Boltzmann. He was an expert in the use of statistical mechanics, but Einstein recognized that statistical mechanics was not the measurement of an individual process, but of multiple processes that could not be measured individually. This may have affected his beliefs about the quantum theory. He regarded the quantum revolution of the 1920s, which depended on probability, as a failure to understand an underlying layer that was deterministic. A single photon cannot interfere with itself. It cannot split and then follow multiple paths.

There is no inherent probability in SUiTe. In that sense, it is an Einsteinian theory. Both de Broglie and Bohm suggested the possibility that particles have guiding waves that are able to go through both slits of the two-slit experiment. The most interesting case is that of the photon since it is itself a wave. It will be necessary to explain precisely what the guiding wave is for a photon and why we don't see it.

With a laser, a photon is emitted by an excited atom through stimulated emission. What if every photon emitted by an excited atom was by stimulated emission? Consider the possibility that space is filled with low-amplitude radiation at all frequencies. The amplitude is too low for individual photons to be measured, but such radiation exists and is found in experiments measuring the Casimir force. Imagine that one of these low-amplitude photons interacts with an exited atom. It has the same frequency and phase as the photon that the excited atom will eventually emit. It stimulates the emission of the measurable photon, but it itself is not measurable. We no longer have a single photon heading toward two slits but two photons.

The sub-threshold photon has two difficulties as a solution for the two-slit experiment. Half of the time it will go through the same slit as the photon that it stimulated. Even when it goes through the opposite slit, it will not be as powerful as a measurable photon going through that slit. The sub-threshold photon helps to explain the fuzzy bands that result from sending photons through one-at-a-time, but not the well-defined bands produced by a laser.

The last necessary piece of the puzzle requires the slits to create waves between the detector and the barrier with the same frequency as the light passing through the slits in the barrier. This differs from the guiding wave theory, in that this wave is created by the slits in response to an electron or a photon passing through the slit. The classical equation for light's behavior in the two-slit experiment treats the slits in exactly this way. The waves generated at the slits guide the photons to their ultimate location at the detector.

This solution suggests that the photon disturbs the electrons of the atoms at the edge of the slit in such a way that they vibrate at the frequency of the passing light. Such a disturbance would be heavily damped. This explanation, combined with the previous explanation, would result in a higher probability of finding a photon in a light band when more photons are passing through the slits per second. When the time between photons is large—say a thousandth of a second—there would no longer be any slit-produced wave, and the chance that a photon will arrive at a light band will depend on the sub-threshold photon passing through the other slit.

In SUiTe, neither a spontaneous emission of a photon that depends solely on probability, nor the similar spontaneous decay of a particle is possible. What appears to be due entirely to inherent probability is due instead to background sub-threshold radiation. The solution given is the only possible one SUiTe can provide for the two-slit experiment. The solution is testable. If all photon pairs are removed—those photons that pass through the two slit experiment consecutively with too little time between them—then the remaining photons should fall into two groups: at least half will create two separate deflection patterns as if the second slit was not there, and the remaining photons will align in the bands. The bands will still be there, but they will be weaker the fewer the number of photons per second. The bands will never disappear because of the sub-threshold photons.

16.3 Bell's Inequalities

Bell responded to an article by Einstein, Podolsky, and Rosen, the famous EPR paper of 1935, by determining the probabilities that both particles in an entangled pair would match the other's spin according to the rules of classical physics or, instead, according to the rules of quantum physics. Quantum physics won.

We can describe Bell's thought experiment in terms of the decay of a neutral pion into an electron and a positron where both particles leave the same location at the same time with opposite momentum, opposite charge, and opposite spin. If I measure the spin of the electron with a Stern-Gerlach apparatus as spin-up, I will know with certainty that the spin of the positron is spin-down.

We can do a thought experiment where the two different Stern-Gerlach apparatuses measure the electron and positron separately. Each of the apparatuses can be rotated to three possible positions: 0 degrees, 120 degrees, and 240 degrees. Classically, if the two apparatuses were separated by enough distance that the speed of light would be too slow to transfer the information on the spin measurement of the electron to the positron before the positron was measured, then the two particles would be unable to coordinate their actions. In quantum theory, the measurement of the spin of the electron instantaneously affects the spin of the positron.

Although Einstein was not aware of the decay modes of the pion—or of the pion itself at the time the EPR paper was written—he did realize that quantum theory would require instantaneous action between particles. He was certain that nothing in nature would allow the passage of information between any two particles faster than the speed of light. He suggested the possibility that there might be some as-yet-undiscovered force in nature that allowed the two particles to appear as if they were still connected. Einstein believed that quantum theory, although correct from an experimental point of view, was an incomplete theory that had ignored an underlying layer in nature capable of replacing inherent probability and instantaneous action at a distance. This missing information came to be called hidden variables. Momentum, electric charge, and spin are also variables that describe a particle, but they are not hidden since they can be directly measured.

Classically, if an electron without any hidden properties is measured as spin-up by one Stern-Gerlach apparatus, and a second apparatus has the same setting (e.g. 120 degrees with 120), then the corresponding positron going through the second apparatus will be measured as spin-down. This would constitute a match in the experiment. In the quantum case where two apparatuses are set to different angles, we could treat the two particles of the Bell thought experiment as if they were a single particle that had had its spin set by one Stern-Gerlach apparatus, and that was then passed through a second apparatus. We would use the formula we previously found for the probability that depended on the difference in angle between the two apparatuses: $(\cos\theta/2)^2$. In the case where both of the apparatuses are set to the same angle, $\theta = 0$ degrees and $(\cos\theta/2)^2 = 1$. The probability of a match is then 1. Both the classical theory and the quantum theory reach the same result.

Classically, if a spin-up electron arrives at an apparatus such that it is within 90 degrees of the magnetic-field direction of the apparatus then it will be measured as spin-up. The corresponding positron arriving at an apparatus rotated 120 degrees with respect to the first will only be measured as spin-down if its spin is between 30 and 90 degrees. For example, if the electron arrived with spin-up at 10 degrees, the positron would arrive with spin-down at 10 degrees at the second apparatus tilted to 120 degrees. It would not be within 90 degrees of the second apparatus's setting for spin-down and would not be measured as spin-down. There would be no match.

Over a 60-degree range, the electron will be recorded as spin-up and the positron as spin-down—a match. Over an additional 60-degree range, the electron will be recorded as spin-down and the positron as spin-up—also a match. One-third of the time we will have matches for classical particles without hidden variables when the two apparatuses are 120 degrees apart. You can see this in the following figure by adding together the cross-hatched area and the blank area. The two areas added together make up one-third of the circle.

The quantum expectation is the application of the formula stated earlier at an angle θ of 120 degrees: $(\cos 120/2)^2 = 0.25$. The final step is to add the numbers up. There are nine possible combinations of settings for the two apparatuses, where either can be arbitrarily set to 0, 120, or 240 degrees. When the two apparatuses are set to the same value, we will expect that all the electrons that are measured as up in one apparatus will match all the positrons that are measured as down in the other apparatus. Matching will be 100 percent.

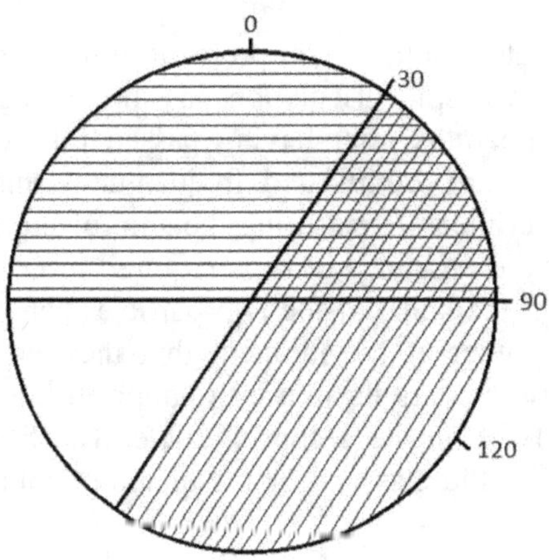

Matching with two Stern-Gerlach apparatuses 120 degrees apart

The classical expectation for a match is one third, the cross-hatched area plus the blank area.

There is a 120-degree difference in the other six possible combinations. Classically, there will be a one-third match, but using the quantum theory we have a one-fourth probability of matching. In six possible orientations of the two apparatuses where they are 120 degrees apart, we have one-third each for classical theory without hidden variables. That is, $6(1/3) = 2$. For quantum theory, we have $6(1/4) = 1.5$. We now add up the three cases where both apparatuses are set to the same value and where both classical and quantum theory agree that the probability of a match will be 100 percent or 1, and we have $3(1) = 3$. Adding together all nine settings for the quantum theory, we have $3 + 1.5 = 4.5$. Adding together all nine settings for the classical theory, we have $3 + 2 = 5$. We now divide by 9 to get the probability of a match for all nine settings of the apparatuses combined. Classical theory requires a match probability of 5/9. Quantum theory requires a match of 0.5. Experimentally, there is a 50 percent probability of a match with these nine combinations—the quantum probability.

The beauty of Bell's argument with these settings is that any hidden variables would only serve to make the classical guess better than 5/9. The only way to make the classical guess worse would be to add a random element into the calculation of the probability of matching. The hidden variables would need to provide precisely what quantum theory supplies: inherent probability.

Although SUiTe does match the quantum probability in the case of a particle with a known spin orientation passing through a second Stern-Gerlach apparatus after having passed through a first, the problem here is more complicated. In quantum terms, in the case of the Bell experiments, measuring the spin of the first particle sets the spin of the second. SUiTe is required to use the classical assumption that the spins of the two particles can only be set relative to each other at the moment that they parted. Measuring the first particle alters its spin by rotating it, but this rotation can have no effect on the second particle. The SUiTe solution will not be either the classical or the quantum solution but a mixture of both.

To simplify the calculation of the probabilities, we will consider the case of two entangled electrons where both have the same spin. A match will be spin-up with spin-up, or spin-down with spin-down. We will consider the probabilities between spin-up or spin-down as independent of each other once they have left the location where they were entangled. We will assume that they share the same spin orientation when they arrive at their respective apparatuses. We will consider the experiment just mentioned, where two apparatuses have the same setting or are set 120 degrees apart. In the two-slit experiment, we found a time dependency that gave a different meaning to the results. It suggested that photons did not interfere with themselves, and that our interpretation of the two-slit experiment was incorrect. We will find that time also plays a role here.

While working out the solution for the Stern-Gerlach experiment, we saw that the electron could immediately become parallel to the threads of the magnetic field, or it could first move back to being perpendicular to the threads before becoming parallel. The probability that it will move directly toward parallel with the threads is $\cos\theta$. If it takes a temporary detour to becoming perpendicular before becoming parallel to the magnetic field, the probability is $0.5(1 - \cos\theta)$. The two probabilities added together give the total probability.

If the two apparatuses have the same orientation, the probabilities should be no different for the second electron entering the second apparatus. In this thought experiment, the requirement for entanglement is that both electrons share the same spin orientation, which for us will be the same angle θ. Because we have assumed that they are independent, we will multiply the probability of one by the probability of the other. This will give us four probabilities for the four different ways in which they could match by both becoming up or both becoming down. We will need to include one last probability for the possibility that after becoming perpendicular to the magnetic field, both electrons continue rotating until they become parallel to the magnetic field pointing down (or up). This last probability is the last line:

$$(\cos\theta)^2$$
$$(\cos\theta)\,0.5(1 - \cos\theta)$$
$$0.5(1 - \cos\theta)\,(\cos\theta)$$
$$0.5(1 - \cos\theta)\,0.5(1 - \cos\theta)$$
$$0.5(1 - \cos\theta)\,0.5(1 - \cos\theta)$$

Added all together, we have $0.5(1 + \cos\theta\cos\theta)$. This creates a problem. Both classical and quantum theory agree that when the two apparatuses are set to the same orientation, the matching will be perfect. The classical probability would be only the first of the five probabilities remembering that if the angle of the incoming electron is within 90 degrees of either up or down, then it will be recorded as up or down. $\cos\theta = 1$ independent of the incoming angle. The quantum theory assumes that two entangled electrons are no different than one electron traveling from one apparatus to another where the first apparatus has set the spin—exactly what we did with the Stern-Gerlach experiment. Again, $\cos\theta = 1$ independent of the incoming angle. In SUiTe, we see that the five probabilities above do not give us perfect matching. In fact, only 75 percent will match; the rest won't.

For apparatuses set 120 degrees apart, the classical assumption is that one-third of the pairs will match. The quantum assumption is that one-fourth of the pairs will match. We will need to introduce another angle into our list of probabilities, which is the absolute value of 120 degrees minus the initial spin angle θ, which we will call ϕ. The probabilities when the two Stern-Gerlach apparatuses are set 120 degrees apart for the range from 90 to 210 degrees are now:

$$-(\boldsymbol{\cos\theta})\,(\boldsymbol{\cos\phi})$$
$$-(\cos\theta)\,0.5(1-\cos\phi)$$
$$0.5(1+\cos\theta)\,(\cos\phi)$$
$$0.5(1+\cos\theta)\,0.5(1-\cos\phi)$$
$$0.5(1+\cos\theta)\,0.5(1-\cos\phi)$$

We can see that the form of the probabilities is nearly identical to what we had before, when both apparatuses were set to the same angle. In the range chosen, $\cos\theta$ is negative, which is why there has been a sign change. The various probabilities add up to $0.5(1-\cos\theta\cos\phi)$. The first probability, $-(\cos\theta)\,(\cos\phi)$, gives a partial probability of the mismatches and not the matches. It has been included to show the similarity with the set of probabilities found when the two apparatuses were set to the same value. The remaining four probabilities give matches.

The total probability of a match can be found by deducting the first probability from the total, which gives us $0.5(1+\cos\theta\cos\phi)$ —exactly the same form as before with the apparatuses set to the same angle, but in this range $\cos\theta$ is always negative.

To complete half of the circle, we will need to add the matching for 30 to 90 degrees to the matching we found with our two apparatuses set 120 degrees apart. Once again we have five probabilities, and they look very much like the probabilities in the range from 90 to 210 degrees, but, this time, the first probability does count toward the matches.

$$(\cos\theta)\,(\cos\phi)$$
$$(\cos\theta)0.5(1-\cos\phi)$$
$$0.5(1-\cos\theta)\,(\cos\phi)$$
$$0.5(1-\cos\theta)\,0.5(1-\cos\phi)$$
$$0.5(1-\cos\theta)\,0.5(1-\cos\phi)$$

In the range of 30 to 90 degrees, neither cosine is negative. The sum is $0.5(1+\cos\theta\cos\phi)$, identical to the form of the probability for matches from 90 to 210, though in this case we include $(\cos\theta)(\cos\phi)$.

$0.5(1+\cos\theta\cos\phi)$ will give answers for entangled particles entering two Stern-Gerlach apparatuses that are 120 degrees apart that are different than either the classical or quantum solutions. The classical value is one-third. The quantum value is one-fourth. The value from SUiTe is 0.375. We need to add up the values for all of the nine orientations of the two apparatuses. For 120 degrees apart that will be $3 \cdot 0.375 = 1.125$. For 240 degrees apart that will be $3 \cdot 0.375 = 1.125$. For the three times that both apparatuses have the same setting, it is $3 \cdot 0.75 = 2.25$. Added together it is 4.5. Divided by 9 it is 0.5, the quantum probability. It is the right answer with the wrong probabilities for each separate piece.

There are three loopholes that can be used to negate Bell's theorems and the experiments that have proven it: distance, time, and sampling error. The distance loophole allows the entangled particles to arrange their spin together by signaling each other at the speed of light. This is irrelevant for SUiTe because the entangled particles are entirely independent of each other after they leave the location where they were entangled.

The sampling error suggests that measuring only a subset of the entangled particles may result in a biased result. This loophole has been tackled by entangling a group of atoms and measuring all of them. The experiment does not allow for measuring individual spin settings as the example given above does. SUiTe passes the sampling test because when all of the entangled pairs are counted, the result is the quantum result.

The final loophole is the time loophole. It is this loophole that may allow the results from SUiTe to give the correct results for the individual settings of the two Stern-Gerlach apparatuses. If we imagine that bumping into the magnetic threads in the Stern-Gerlach inhomogeneous magnetic field not only modifies the spin of the electron but can also slow it down, then those pairs of entangled particles where one particle follows a different path through the Stern-Gerlach apparatus than the other may result in one of the entangled particles arriving at the detector later than the other. The time loophole is necessary because the only way of determining if two particles are entangled, in those experiments where not all particles are counted, is by shortening the time interval to the point that no unentangled particles are counted.

Here is the problem. In both quantum and classical tests of the Bell theorem, two Stern-Gerlach apparatuses set at the same angle should give the same result according to theory. Matching should be 100 percent. If it isn't, this suggests that the time interval is too long. Looking at SUiTe, if the bumping up against the magnetic threads of the inhomogeneous field slows the electron, it is necessary to look at those particles that follow a similar path for the same distance. For two apparatuses at the same setting, this is equivalent to looking only at those particles arriving within 30 degrees or less than the point where both will be recorded as up, or where both will be recorded as down. At least two-thirds of the entangled pairs will be thrown out. The time window must be made very narrow. The closer the experimenter wants the value to be to 1, the more pairs that will be thrown out and the narrower that the time window will be. At only 15 degrees from being measured up or down, with five-sixths of the pairs tossed out, the resulting measurement is close enough to 1 to be acceptable.

Applying this extremely short time window to the apparatuses set 120 degrees apart will center them very near to 60 degrees (up) and 250 degrees (down). But 60 degrees up is 0.25 and 250 degrees down is also 0.25. That is what is expected in quantum theory. A slight increase in the time window will actually lower the total value after all the individual settings have been added up, pushing it away from the classical value. The time loophole can bias the experimenter toward the expected quantum result by adjustments to the time window. The Bell tests were ultimately theory dependent and not observation dependent. A different theory would have led to different conclusions.

The most important result of this section is the matching of the SUiTe result with the measured result when all entangled pairs are measured. In the standard model, this is due to the assumptions made in the present quantum theory. These assumptions depend on various principles that do not match with classical physics. By turning the electron into a classical particle and describing, in a classical way, how it interacts with an inhomogeneous magnetic field, it was possible to obtain the same results without any classically unsupported principles or hidden variables. The Bell theorem and test experiments are pillars of modern quantum theory. If in fact the experiments can be described classically, then Einstein was right. There is no "spooky action at a distance." All physics is local.

Chapter 17

Relativistic Wave Equation

The basis for the various relativistic wave equations is the relativistic energy-momentum relation

$$E^2 - m_r^2c^4 = p^2c^2$$

where m_r is the rest mass. Going directly from this relation to a wave equation leads to the Klein-Gordon equation. The Klein-Gordon equation has negative energy and negative probability solutions. There is no reasonable interpretation of the Klein-Gordon equation which allows for negative probabilities. Dirac factored the energy-momentum relation into two solutions that were first-order rather than second-order. This removed the negative probability solution but not the negative energy solution. The Dirac equation also has a solution when the velocity is zero and the momentum is therefore zero.

The zero-velocity solution to the Dirac equation is a bispinor—a 4×1 vector—where the first two rows are a spinor with positive energy, the electron, and the last two rows are a spinor with negative energy, the positron. The negative energy is interpreted as being the positive energy of antimatter. In addition to the energy, the spinor also describes the spin of the particle. The usual interpretation of the Dirac zero-energy solution suggests that, with sufficient energy, two particles could be made—matter and antimatter—with the same mass and opposite spin.

Both the Klein-Gordon and Dirac equations represent problems for SUiTe. Positrons, electrons, and other particles cannot be created directly from energy. Probabilities cannot be negative and the only negative energy is binding energy. Even the name antimatter, though I use it for convenience, has no sense in SUiTe. A positron is identical to an electron except for its electrostatic charge. The Dirac equation is not necessary to provide evidence for antimatter. Another problem with the antimatter of the Dirac equation is that it suggests that combining an electron and a positron will leave nothing but energy. Combining the two does release electrostatic potential energy as they come close to each other, but the electron and positron do not disappear; instead, they become two sides of a particle of dark matter.

Spin is also different. Both the electron and the positron have three spin states and not two. The most important of the three states is when one electron sets its spin perpendicular to that of another in an electron shell. The remaining two states are those that can be seen in a Stern-Gerlach apparatus. The spin measured in this way is not an intrinsic property but can be easily changed by a magnetic field if it generates a rotation of the electron about its electrostatic forceline.

17.1 The SUiTe Relativistic Wave Equation

The Schroedinger equation is

$$i\hbar \, {}^{\partial\psi}\!/_{\partial t} = -{}^{\hbar^2}\!/_{2m} \nabla^2\psi + V\psi$$

The original Dirac equation is

$$i\hbar \, {}^{\partial\psi}\!/_{\partial t} = \left(\beta mc^2 + c\left(\sum_{n=0}^{3} \alpha_n p_n)\right)\right)\psi + V\psi$$

This can be rewritten as

$$i\hbar \, {}^{\partial\psi}\!/_{\partial t} = \beta mc^2 - i\hbar c(\boldsymbol{\alpha} \cdot \nabla\psi) + V\psi$$

Both the Dirac and Schroedinger equations are first order in time. This is necessary if the equation is to be used to determine the energy levels in the hydrogen atom. It is also necessary for the Born rule. The Born rule interprets the wave function of either the Schroedinger or Dirac equations as a probability amplitude. Squaring the wave function gives the probability of finding a particle in the relevant volume.

To find a relativistic wave equation that is similar to the Schroedinger equation, we start with the energy momentum relation: $E^2 - m_r^2c^4 = p^2c^2$. The simplest way to proceed from the relativistic energy-momentum relation to a first-order wave equation is to note that the second term on the left does not depend on velocity and is always less than the first term on the left for a non-zero velocity. We will assume that the rest mass of the electron is a constant. We can introduce a constant k such that $0 \leq k < 1$ and $k^2E^2 = E^2 - m_r^2c^4$. Then the relativistic energy-momentum relation becomes: $k^2E^2 = p^2c^2$ or $k^2m^2c^4 = m^2v^2c^2$ and $k = {}^{v}/_{c}$ where m and v are the relativistic mass and the velocity.

The resulting wave equation (RWE) is:

$$i\hbar\,{}^{\partial\psi}/_{\partial t} = -i\hbar c({}^{c}/_{v})\nabla\psi + V\psi$$

The left-hand side is the same as in the Schroedinger and Dirac equations. The first term on the right, without the value in parentheses, is what the Dirac term would look like if there was no rest energy in the Dirac equation and if the Dirac matrices were all equal to 1. In SUiTe, the particle's wave reflects the difficulty of deflecting the particle. As the velocity of the particle increases, the wavelength decreases and the deflection also decreases.

In the Schroedinger equation, a free electron is spread throughout space instantaneously. In the RWE, the wave expands at the rate of c^2/v. Only an electron at rest is able to reach out throughout space as the electron does in the Schroedinger equation. The faster the electron moves, the shorter the distance over which it affects the universe-wide electrostatic field and the less it is affected by that field. As an electron approaches the speed of light, it becomes light-like.

In SUiTe the longitudinal compression of the w-duals of space causes the transverse compression to spread outward at the speed of light and then collapse at the speed of light as the photon passes (section 10.1). A massive particle traveling at the speed of light is no different than a photon in terms of the spreading of the wave, as can be seen from the above equation. In that case, the physical part of the wave function is limited to the distance determined by the speed of light in the time it takes for the particle to pass a specific location.

17.2 Hydrogen Energy Levels

The Schroedinger equation is second-order in *x*, so there are two solutions when we wish to determine the energy of a wave-particle confined to a box. De Broglie determined that the momentum of the orbiting electron in the rest state of a hydrogen atom had a wavelength equal to the circumference of a sphere that had a radius equal to the Bohr radius. We can think of this circumference as a narrow box $2\pi r_B$ long. The two solutions for the wave function for the particle in a box are:

$$\psi_1(x) = A \sin kx \text{ and } \psi_2(x) = B \cos kx$$

The second solution is thrown out because when $x = 0$, at the side of the box where the wave begins or ends, the wave function also has to be 0. Solving the first of these two equations for *k*, we have $k = {}^{n\pi}/_{2\pi r}$.

A problem with attempting to determine the energy levels from the Schroedinger equation in this way is that *r*, in the denominator, must be multiplied by the square of the shell number. In the Bohr model, the distance to the first shell is about 52.9 picometers and the distance to the second shell is four times larger. To the third shell the distance is nine times larger than to the first. Each energy shell represents a new box in which the particle wave of the electron is constrained. We are not interested here in *n* different solutions for each energy shell, but only in one solution. *n* in the above equation must always be equal to 1 whereas *r* changes. In SUiTe, *r* is an integer multiple of one-half of the Bohr radius, r_h, the minimum covalent radius of hydrogen, and therefore, $k_n = {}^{\pi}/_{2\pi n r_h}$. With these changes, we can determine the energy levels using the RWE.

The energy found by the Schroedinger equation at the first shell of hydrogen can be better described as the kinetic energy of an electron orbiting the nucleus at a speed of αc. Using de Broglie's relation between momentum and wavelength, $\lambda = {}^{h}/_{p}$, and momentum equal to $m_e\alpha c$, we find the wavelength of $2\pi r_B$. The kinetic energy of the orbiting electron is half the electrostatic potential energy at a distance from the nucleus of r_B. The other half of the energy is emitted as a photon with an energy of about 13.6 *eV*. The Bohr radius was chosen so that the energy of the emitted light plus the kinetic energy of the orbiting electron would sum to the electrostatic potential of about 27.2 *eV*.

The RWE in one dimension is:

$$i\hbar\, {}^{\partial\psi}/_{\partial t} = -({}^{c}/_{v})i\hbar c\, {}^{\partial\psi}/_{\partial x}$$

when we set changes to the potential, *V*, to 0, which it is inside a narrow box at an unchanging distance from the nucleus. This can be rewritten as $E = \hbar c({}^{c}/_{v})\mathrm{k}$. k_n is the same as before: $k_n = {}^{\pi}/_{2\pi n r_h}$ where n in the numerator has been set to 1 and the radius is the hydrogen covalent radius. We can substitute for *k* and solve for *E* when $n = 1$:

$$E = {}^{(\pi\hbar c^2)}/_{(2\pi r_h v)}$$

We can simplify further because $r_h = {}^{\hbar}/_{2m_e c\alpha}$ and then

$$E = m_{re}c^2\alpha({}^{c}/_{v})$$

Both *E* and $m_{re}c^2$ are the relativistic energy and must be equal, so $\alpha({}^{c}/_{v}) = 1$ and $v = \alpha c$ where α is the fine-structure constant. But the relativistic energy includes the rest energy. We want the available energy which is the relativistic kinetic energy. Knowing the velocity, we can compute the available energy by using the rest mass of the electron rather than the relativistic mass and multiplying by the Lorentz factor minus 1.

$$\left({}^{1}/_{\sqrt{1-\alpha^2}} - 1\right)E = m_e c^2\left({}^{1}/_{\sqrt{1-\alpha^2}} - 1\right) = 13.6.\, eV$$

This is the energy of the emitted photon. The only energy available for the emitted light must be the relativistic energy minus the rest energy. This reflects the fact that the electron has come to a stop at the first shell, and its forcelines are no longer tilted. To save space in the following equations I will introduce the Lorentz factor, γ_L.

$$\left(\sqrt{1 - v^2/c^2}\right)^{-1} - 1 = \gamma_L - 1$$

The non-relativistic kinetic energy is:

$$p^2\Big/_{2m_e} = {}^{m_e}\!/_2\,(\alpha c)^2 \cong 13.6\,eV$$

The speed of the electron, as it brakes to a halt at the first shell, emitting a photon as bremsstrahlung, is less than 1 percent of the speed of light, so that the non-relativistic kinetic energy is a good approximation for the relativistic kinetic energy.

As we look for the energy at shells farther away from the nucleus, we need to know the velocity of the electron at the point that it reached the shell from which the photon was emitted. Even though we don't know the exact velocity, we can set $v_n \cong {}^{v_1}\!/_n$ because $v_1 = \alpha c$ is small compared to the speed of light. Using ${}^{v_1}\!/_n$ for the velocity of the nth shell, the total energy for the nth shell is:

$$(\gamma_L - 1)E_n = \frac{\pi\hbar c^2}{n2\pi r_h\left({}^{v_1}\!/_n\right)} = (\gamma_L - 1)\frac{\hbar c}{2r_h\alpha} = (\gamma_L - 1)m_e c^2$$

To compute the available energy for the nth shell:

$$A_n \cong \frac{\hbar c}{2r_h\alpha}\left(1\Bigg/\sqrt{1 - \alpha^2\!/_{n^2}} - 1\right) \cong \frac{1}{2n^2}m_e(\alpha c)^2$$

Where the term in parenthesis is the Lorentz factor minus 1, $\gamma_L - 1$. This result requires two approximations. Referring back to section 2.1 on special relativity, the denominator in parenthesis is $\sin(\cos^{-1} v/c) = \sqrt{1-(v/c)^2} \cong 1 - 0.5(v/c)^2$ from the Taylor expansion when $(v/c)^2 \ll 1$. There is a final additional approximation that $1/(1-x) \cong 1 + x$ for $x \ll 1$.

Finding the division by n^2 is important. There were three problems with the combination of the Bohr atom with the Schroedinger equation: (1) multiplying the square of the shell number by the Bohr radius created absurdly large excited hydrogen atoms, (2) the electron was required to orbit the proton in the hydrogen atom even though it is known that an electron would rapidly lose energy and spiral into the nucleus, (3) the assumed jumps of the electron between shells occur outside of space and time. Here, we can see that the energy emitted is equal to the kinetic energy lost by the electron as it comes to a stop. The need for the square of the shell number is no longer ad hoc, but is an essential requirement of replacing the Schroedinger equation with the RWE. It comes from special relativity.

Since we will only be interested in the available energy (relativistic energy minus the rest energy) when using this wave equation in place of the Schroedinger equation for determining energy levels, the relativistic wave equation should be written as:

$$(\gamma_L - 1)i\hbar\, {}^{\partial\psi}/_{\partial t} = (\gamma_L - 1)(-({}^{c}/_{v})i\hbar c\nabla\psi + V\psi)$$

For the hydrogen energy levels, the Schroedinger equation gives the right answers for the wrong reasons. The energy emitted by an electron as it reaches a specific shell is what we would like to calculate, not the electron kinetic energy as it orbits the nucleus at the Bohr radius times the square of the shell number.

17.3 Is the RWE Really a Wave Equation?

Like the Dirac wave equation, RWE is the square root of the wave equation. For a free particle (a particle without potential energy), after removing shared terms from both sides, the RWE is

$${}^{\partial\psi}/_{\partial t} = -({}^{c}/_{v})c\nabla\psi$$

Squaring both sides and moving terms we have

$$\partial^2\psi/_{\partial t^2}\left(1/_{c^2}\right) = (c^2/_{v^2})\nabla^2\psi$$

Without c^2/v^2, which is a constant when a particle is not subject to a field, the above equation is the wave equation of Maxwell for a particle of light in space. We see again, that as the speed of a particle approaches the speed of light it becomes light-like, and the wave equation for light and for the particle become identical.

17.4 Calculating the Wave Number

To determine transmission or reflection at a boundary created by a square wave potential, or to determine the probability of locating a particle using the square of the wave function, it is necessary to calculate the wave number. The wave number for a plane wave is $2\pi/_\lambda$ where λ is the wavelength. The value of the wave function depends on both the time and the displacement from some point in space and time where the wave originated. The wave function provides the probability amplitude. This value, when squared, is the probability of finding a particle in the relevant volume according to the Born rule. As Planck found in his solution to the blackbody radiation problem in 1900, the amplitude of the wave of a photon is irrelevant in determining the energy of a photon. The amplitude of the wave function had to take on a different role in quantum physics: the square root of a probability.

In the Schroedinger equation, for a square wave potential, the wave number, *k*, can be determined from:

$$k_S^2 = {}^{2m_e(E_S - V)}/_{\hbar^2}$$

E_S is the Schroedinger equation energy and is equal to the non-relativistic kinetic energy. m_e is the rest mass of the electron. *V* is the potential energy that must be overcome. The wave number for the relativistic wave equation is:

$$k_{rwe} = 2(^c/_v)\,{}^{(A-V)}/_{\hbar c}$$

where A is the relativistic kinetic energy and is equal to the available energy. If we only have the kinetic energies and potentials for non-relativistic particles, we can use them in place of the relativistic kinetic energy: $E_S - V \cong A - V$. We can calculate the relative velocity from the equation below, where m_r is the rest mass of the electron, if we assume that $A - V$ is the available relativistic kinetic energy. The following equation follows from the definition of the relativistic kinetic energy and that the Lorentz factor is equal to $(\sin\theta)^{-1}$ where $\theta = \cos^{-1}(^v/_c)$. The important thing to remember is that, in SUiTe, the rest mass energy is not available to be used. Only the relativistic kinetic energy is available. The potential energy can never be deducted from the rest mass energy.

$$(^v/_c) = cos\left(\sin^{-1}\left(1 + {}^{(A-V)}/_{m_r c^2}\right)^{-1}\right)$$

Let's do an example where $E_S - V = 100eV \cong A - V$. Using the two equations above, $^v/_c$=0.01978 and $k_{rwe} = 5.12 \times 10^{10}$. For the Schroedinger equation wave number, $k_S^2 = 2.626 \times 10^{21}$ and $k_S = 5.12 \times 10^{10}$. As was the case with the hydrogen energy levels, the answers match.

Chapter 18

Nuclear Structure and Magic Numbers

If the atomic nucleus is a sphere filled with balls that are all the same size, the radius should be roughly equal to $r = A^{1/3} \cdot R$ femtometers where A is the number of nucleons, and $1 \leq R \leq 1.2$. The problem with simply filling a sphere is that the elemental isotopes follow various rules that depend on a specific arrangement of the nucleons in the nucleus.

There are certain elements that appear to be landmarks of some sort. Calcium is the last element that has a stable isotope with an even number of protons and neutrons. Lead is the largest stable element, with 82 protons. Tin has more stable isotopes than any other element. Could the special status of calcium, tin, and lead be found if nature arranged the nucleons of the nucleus in layers?

If we take the above equation for the radius of a nucleus seriously and, at the same time, add in balls until a layer is filled, we should be able to find the element that corresponds to each level with all layers filled. These layers are the *shells* of the nuclear shell model. The first shell must surround a sphere filled with nucleons. The nuclei of atoms are cannot be hollow in the center. The rules that govern the shells are different than the rules that govern the first seven elements. The first shell starts at oxygen with 8 protons and 8 neutrons.

We can make a simple ansatz that suggests numbers for nuclides one shell apart. We can start with oxygen where the number of nucleons is roughly equal to 2.5 cubed. If shells are one nucleon thick, the next shell should be at 3.5 cubed, then 4.5 cubed, and so on. The number of nucleons we can put in a nucleus for the initial filled sphere is 2.5 cubed (15.6). We then round to the nearest even number and find the closest stable isotope that has that number of nucleons, oxygen 16. The second would be 3.5 cubed (42.9). That would correspond to the element calcium.

Level	Value	Element	Protons	Isotope
1	15.6	Oxygen	8	O 16
2	42.9	Calcium	20	Ca 43
3	91.1	Zirconium	40	Zr 92
4	166.4	Erbium	68	Er 166

We can see that the sequence does include calcium but does not include tin or lead. We can solve for lead and then subtract 1 for each lower shell. The most common isotope of lead has 208 nucleons, which is 5.925 cubed.

Shell	Value	Element	Protons	Isotope
4	208.0	Lead	82	Pb 208
3	119.5	Tin	50	Sn 120
2	62.5	Nickel	28	Ni 62
1	25.0	Magnesium	12	Mg 24

The first two numbers in the proton column in both tables are called magic numbers because isotopes of elements that have either that many protons or that many neutrons appear to be more stable than other nearby isotopes. At 208 nucleons, lead is considered to be doubly magic because both 82 (protons) and 126 (neutrons) are believed to be magic numbers. Oxygen 16 and Calcium 40 are also doubly magic.

In both tables, the problem occurs at the third row. The distance between calcium and tin is too large. Perhaps this is because elements are created by a fusion chain process through nickel 56. The creation of these elements releases energy. After nickel 56, elements are mostly created by adding single neutrons.

18.1 Nuclear Shell Model

Mayer, Wigner, and Jensen received the Nobel Prize in 1963 for the development of a model of the nucleus composed of shells that when the shells were full and added together, corresponded to those numbers considered to be magic at that time. The solution of the problem of the third shell (the fourth and fifth shells for Mayer et al.) suggested that the initial shell was 2 (helium), and it combined with a subshell of 6 in order to match the magic number of 8 for oxygen, the second shell. The third subshell became 12, and with the addition of the 2 shell and the 6 subshell was 20 calcium, the third proton-magic element of their theory. The fourth shell of the Mayer model was nickel which was calcium with the 2 shell and the 6 subshell added. Nickel then became magic with 28 protons. Tin required 22 more protons and lead, 32 more protons.

Here is a comparison of the shell and subshell structure of the electron shells with the atomic nuclei of the above model. For the electrons associated with the noble gases after helium, the *s* and *p* shells must be full. Neon is s:s,p; argon is *s:s,p:s,p*; krypton is *s:s,p:s,p,d:s,p*; xenon is *s:s,p:s,p,d:s,p,d:s,p*; and radon is *s:s,p:s,p,d:s,p,d,f:s,p,d:s,p*. In the Mayer nuclear shell model, the first shell corresponds to $s = 2$, but the second shell has a single subshell with $p = 6$, and the total for both of the first two shells is then 8, and not 10 as we see with electrons in neon, the second of the noble gases. After helium we have *s:p* for oxygen, perhaps *s:p:s,d* for calcium, *s:p:s,d:s,p* for nickel, *s:p:s,d:s,p,d:s,d* for tin, and finally *s:p:s,d:s,p,d:s,d:s,p,d,f* for lead. The comparison of atomic nuclei with the electron configurations of the noble gases was a poor choice.

An important problem with the old shell model is that nickel isn't magic. If it had been proton-magic, ${}^{66}_{28}Ni$ would be stable just as ${}^{46}_{20}Ca$ is stable and ${}^{124}_{50}Sn$ is stable. Both calcium and tin are able to expand beyond their normal range to the right (neutron-rich side) by one nucleon. Nickel can't. While 8, 20, 50, 82, and 126 can be shown to be magic for neutrons, and the first four are magic for protons, 28 depends entirely on a theory which has shown itself to be surprisingly barren.

There is another problem with the old nuclear shell model. At the start of the chapter, I was able to find all but one of the magic numbers, the number 126, by filling a sphere, starting either from 2.5 going up from oxygen 16 or 5.925 going down from lead 208, adding or subtracting 1, respectively. The measured size of these nuclei is close to what would be expected from the formula at the beginning of this chapter. The nuclear shell model of Mayer, Wigner, and Jensen requires separate shells for protons and neutrons. There isn't enough space in a nucleus to accommodate their model.

18.2 The New Shell Model

A new shell model must offer some advantages compared to previous models of atomic nuclei. The previous shell model provided magic numbers, not always correct ones, but little else. The liquid drop model gives accurate estimates of the diameter of the various nuclides and associates a given nuclide with its most likely element. Calculating the center of the valley of stability for proton numbers successfully yields at least one stable isotope for most elements but fails to explain the lack of a stable isotope for technetium or promethium, nor can it explain why there are no stable nuclides with numbers 147 and 151.

A new model should ideally provide the magic numbers, provide reasonable estimates of the diameters of the nuclei, and associate stable nuclides with specific elements. A new model should also explain what causes some isotopes to be unstable even when they are within the usual range of nuclides that should be stable for a particular element given its proton number.

We can start by defining the shells. The old model has separate shells for neutrons and protons; the new model combines them. The first listed shell for the new model will be oxygen which is actually a sphere full of nucleons and not a shell. The first true shell is constructed around this oxygen core. This is the calcium shell which, when full, results in the double-magic $^{40}_{20}Ca$. Tin is proton-magic with 50 protons, but even at $^{86}_{36}Kr$ all the places available to neutrons in the tin shell have been filled. The lead shell runs out of places for neutrons at $^{138}_{56}Ba$ even though there are still places for 26 additional protons. An atomic shell can completely fill with neutrons without protons. This is the case with $^{208}_{82}Pb$ with 44 neutrons and no protons in the outer shell.

Protons and neutrons can bind with each other in the same shell using the weak force. Two neutrons can also bind together using the weak force. Two deuterons can bind together to form an alpha particle on the shell surface, and four neutrons can also bind together to form a similar but weaker and larger version of an alpha particle. The after-lead shell is filled with only neutrons. This is the case with lead 208, where the lead shell of 64 nucleons—half protons and half neutrons—that begins its construction with zirconium 91, is covered by the after-lead shell using 44 neutrons and no protons.

It is necessary to put as many nucleons into as small a space as possible. This can be done if we alternate rows of neutrons and protons within a shell. We will always have an odd number of rows: 7 rows for the implied first shell at ${}^{16}_{8}O$, 15 rows for the fifth shell filled with only neutrons at ${}^{208}_{82}Pb$. Nature avoids placing a proton on top of a proton or a neutron on top of a neutron in adjacent shells. This forces the largest row, the one along the equator, to switch from protons to neutrons in adjacent shells. The binding between adjacent shells is not due to the weak force that binds nucleons together on the shell surface but is due to the electrostatic force between the shells.

Most full shells have equal numbers of protons and neutrons. The fifth shell appears to have 2 fewer protons than neutrons. The sixth shell (not shown) should then have 2 fewer neutrons than protons. The total number of protons or neutrons in a shell must be an even number. In the following table, proton numbers are in bold italics. With the exception of the first line, which refers to the filled sphere of nucleons that corresponds to oxygen 16, each line in the following table represents an entire spherical shell. The center number in the line is the number of either neutrons or protons along the equator. The columns represent the number of neutrons or protons in the rows of a shell, starting and ending with the poles.

Nuclear Shells

															rows	p+n
				1	2	***3***	4	***3***	2	***1***					7	16
			0	2	***3***	4	***6***	4	***3***	2	***0***				7(9)	24
	0	***2***	3	***4***	6	***9***	12	***9***	6	***4***	3	***2***	0		11(13)	60
	1	2	***3***	4	***6***	10	***12***	10	***6***	4	***3***	2	***1***		13	64
0	2	***3***	4	***6***	9	***12***	14	***12***	9	***6***	4	***3***	2	***0***	15	86

It is not possible to have protons on top of protons in adjacent shells at the poles—hence the zeros in the second and fifth lines.

If we leave out the problematic third line, corresponding to the fourth and fifth shells in the old version of the nuclear shell model, there is a clear pattern in the numbers, but there is also a problem. In the last and fifth shell, it is necessary to put protons at the poles so that the proton and neutron numbers are equal, yet the previous shell also has protons at the poles. The last shell must then have 44 neutrons, which we can verify by showing that 126 neutrons is a magic number, but the next proton-magic-number after lead will be element 124 and not element 126. The next double-magic element should then be ${}^{308}_{124}Ubq$.

If protons and neutrons are reversed in the first shell, they must be reversed in all subsequent shells. There is another potential path that starts with protons at the equator in the oxygen shell rather than neutrons. Only the one shown above leads to ${}^{208}_{82}Pb$. Although I can see no reason why protons and neutrons couldn't be reversed, my past experience with SUiTe has taught me that there are no arbitrary decisions.

From the table above, we can find magic numbers at 8, 20, 50, 82, and 126 (124 for protons). When we examine the isotopes more carefully in the next section, we will find that those five are the only true magic numbers for neutrons and protons. Nickel is an important number as it signals the end of the fusion elements, but it is not proton-magic.

18.3 Determining the Elemental Isotopes

By interleaving neutrons and protons in the same shell and placing one shell directly on top of the other, we can make a shell model that will produce approximately the same diameters for nuclei as we can make with either random close packing or with the liquid drop model. What we would like to have now is a means of determining all the stable isotopes for every element.

In order to find the stable isotopes for an element with an even number of protons, we must first determine the start and end of the range for that element. The range, R, increases from $R_0 = 2$ to $R_1 = 3$ with oxygen. Adding 8 to the number of protons in oxygen reaches sulfur where $R_2 = 5$. Sulfur is also the first element to have filled the oxygen shell (the shell around oxygen) with neutrons. Add 12 protons to sulfur to reach nickel where $R_3 = 7$. Add 16 protons to nickel to reach ruthenium where $R_4 = 9$ which is also the maximum range.

Small adjustments to the range are necessary at regular intervals. The right-hand-side (neutron-rich side) must be reduced by 1 neutron resulting in the exclusion of ${}^{76}_{32}Ge$, ${}^{128}_{52}Te$, and ${}^{182}_{72}Hf$ from the list of stable isotopes. These three elements with a reduced range are found using the equations $8R_1 + 8 = 32$, $8R_2 + 12 = 52$, and $8R_3 + 16 = 72$. Again, we see the sequence of adding 8, 12, and 16 that was used in determining the ranges. There is a second, even smaller adjustment of $2(8R_1 + 8) = 64$ where the range is increased by one neutron allowing ${}^{160}_{64}Gd$.

The magic numbers of 8, 20, 50, 82, and 126 correspond to the beginning of the oxygen shell, the calcium shell, the tin shell, the lead shell, and after-lead shell. The two proton-magic elements calcium and tin are able to extend their range to the right (neutron-rich side) by one nucleon. This is the case with ${}^{46}_{20}Ca$ and ${}^{124}_{50}Sn$. An even element range can stretch by one nucleon to the left (proton-rich side) if that will allow it to have a neutron-magic isotope. This only occurs with ${}^{144}_{62}Sm$. In the case where an even element has a neutron-magic isotope within the range, the range is truncated after the neutron-magic isotope on the side away from the midpoint. There are two elements where the neutron-magic isotope appears to land on the midpoint, sulfur and cerium, in which case neither side is truncated.

The isotope range for odd elements is always within $7/6$ of the midpoint. If the midpoint is within 1/6 of an even number the odd element will have two isotopes. Beginning with oxygen, the even-element isotope range is 3. The midpoint value for oxygen is 16.5. The endpoints of the range for even elements must be even isotopes. Although the range for oxygen is from 15 to 18, the possible isotopes are then 16, 17, and 18. Odd nuclides not claimed by two adjacent odd elements belong to the even element in between unless an adjacent even element has the same nuclide closer to the midpoint. For example, the stable isotopes of ${}^{15}_{7}N$ and ${}^{19}_{9}F$ leave nuclide 17 available for oxygen.

There is an ordering when associating an odd nuclide to an element. While even nuclides may be associated with two different elements, with only one exception, odd nuclides have only one element. The claim of an odd element over a nuclide comes first. If it is not in the range of an odd element, it belongs to the even element whose odd nuclide is closest to the midpoint. But there is a rule as important as the requirement that an odd nuclide within 7/6 of the midpoint of an odd element is an isotope of that element. After nickel, if an odd nuclide is within 7/6 of the midpoint of an even element, it is an isotope of that even element. What happens when both rules apply? The isotope then belongs to both an even and an odd element. ${}^{123}_{52}Te$ and ${}^{123}_{51}Sb$ represent the only such shared odd nuclide.

An important adjustment not associated with the range or magic numbers is the tightening of the nucleus starting at iron which lasts until copper associated with the end of the fusion sequence. This pulls the midpoint toward the left from its estimated value because the protons in the four elements affected do not need to have as many neutrons as expected in order to balance the increasing positivity of the nucleus.

The remaining adjustments to the isotope list are due to instabilities in the shells. The most important instability occurs as the first neutrons are being added to a new shell. This instability happens with 4 neutrons and 6 neutrons but can include an even element nuclide with 2 neutrons if it is involved in an alpha decay cascade as is the case in the lead shell that begins with 83 neutrons. This instability primarily affects the odd elements, especially odd elements with two isotopes. There is no instability in the calcium shell of any type, but the instabilities get worse as the shells get larger with some even element isotopes being affected in the lead shell.

Without this instability, starting after the calcium shell begins to form, ${}^{49}_{23}V$ should be stable, but with 6 neutrons, it fails. ${}^{51}_{23}V$, with 8 neutrons, is stable. In the beginning of the tin shell, both potential isotopes of technetium fail, ${}^{97}_{43}Tc$ with 4 neutrons in the new shell and ${}^{99}_{43}Tc$ with 6 neutrons. These three nuclides appear as stable isotopes in the surrounding even elements. This is not the case in the lead shell where the two nuclides affected, ${}^{147}_{61}Pm$ with 4 neutrons and ${}^{151}_{63}Eu$ with 6 neutrons, have no corresponding stable even element isotopes at 147 or 151.

The remaining instabilities occur only in the lead shell and are notable for using alpha-decay to reach stability. The instability associated with having 2, 4 or 6 neutrons in the lead shell for an even element depends on finding a path to either ${}^{140}_{58}Ce$ or ${}^{142}_{8}Nd$, both of which are neutron-magic. ${}^{152}_{64}Gd$, with 6 neutrons, alpha-decays to ${}^{148}_{62}Sm$ with 4 neutrons, which then decays into ${}^{144}_{60}Nd$ with 2 neutrons, before reaching ${}^{140}_{58}Ce$, an isotope that is neutron-magic and stable. There is a similar path to the neutron-magic ${}^{142}_{60}Nd$ from ${}^{150}_{64}Gd$ with 4 neutrons via ${}^{146}_{62}Sm$ with 2 neutrons. Although ${}^{144}_{58}Ce$ beta-decays twice to ${}^{144}_{60}Nd$, ${}^{146}_{60}Nd$ does not beta-decay twice to reach ${}^{146}_{62}Sm$, an isotope which alpha-decays. A stable ${}^{144}_{58}Ce$ is an error.

The last instability, which only appears in the lead shell, is centered around tungsten. Like the instability of the previous paragraph, it is necessary for an unstable isotope to alpha decay into a stable isotope. There appear to be two factors at play, one is that ${}^{180}_{74}W$, with 24 neutrons, is just past the midpoint of the 44 neutrons needed to fill the shell. The other factor is the ratio of the number of protons in the tin shell divided by the number of neutrons in the lead shell. The ratio is 1 at ${}^{180}_{74}W$. Only the most proton-rich isotopes of the even element before tungsten, ${}^{174}_{72}Hf$, and the even element after, ${}^{186}_{76}Os$, are affected.

The alpha particles that are emitted in the instabilities of the lead shell do not appear to form on the shell surface, but include the calcium, tin, and lead shells, but not the oxygen shell, pulling out a proton from the calcium and lead shells along with two neutrons from the tin shell. It may be triggered when a neutron in the lead shell beta-decays into a proton.

The Equation for the Midpoint of the Range

$$A = 2Z + a\,(Z - N_m)/2 + b\,(Z - N_m)/6 + x$$

where A is the midpoint, Z is the number of protons, N_m is the magic number, a and b are either 1 or 0, and x, at least until terbium, depends on the magic numbers. A corresponds to the center of the valley of stability.

Changes to a and b occur at proton magic numbers (oxygen and tin) or near them (argon), and changes to x occur at either proton or neutron magic numbers through cerium but not after. When the endpoint of one group of elements equals the starting point of the next, the equation must be evaluated using both the earlier and later values of the constants. The midpoint for the element which is at the change point is the average of the two values. This is not the case for the change point at tin which is the only change point at a proton-magic element. For tin, both values are taken as midpoints and those isotopes that fall in between the midpoints are stable for tin.

The greater instability of the lead shell is reflected in the many change points, most of which are not at magic numbers. Odd and even elements no longer have the same midpoints from tin until thulium. It isn't until osmium that it is again possible to take the average between the previous and the next values of the midpoint equation. Even with the many changes to x, the midpoint for lanthanum given by the above equation is too low.

Starting at osmium, the proton-neutron ratio is too high for stability, and it is necessary to add an additional neutron. x shifts from 19 to 20. There is a similar additional shift at thallium of $8/3$ neutrons where x shifts from 20 to 22.67. Neither of these shifts has anything to do with the shell theory except that without the shifts, it would not be possible to fill the neutron portion of the lead shell in time for the double-magic $^{208}_{82}Pb$. At $^{204}_{80}Hg$ (mercury), the proton-neutron ratio is at its lowest. It is no longer possible, given the constraints of having nearly equal numbers of protons and neutrons in shells, for lead to have an isotope with a smaller ratio without adding a neutron shell on top of a neutron shell. This is not possible because neutrons are unable to attach to other neutrons between shells. Neutrons must attach to protons in the shell below. Beyond lead, up to two neutrons can attach to a proton that enters into the after-lead shell.

The variables a, b, x, and N_m change as follows:

8-18*	both	$a = 0, b = 1, N_m = 8$	$x = 0.5$
18-39*	both	$a = 1, b = 0, N_m = 20$	$x = 2.5$
39-50*	both	$a = 1, b = 0, N_m = 20$	$x = 0.5$
50-58*	even	$a = 1, b = 1, N_m = 50$	$x = 18 + 5/6$
51-57	odd	$a = 1, b = 1, N_m = 50$	$x = 19 + 1/3$
60-68	even	$a = 1, b = 1, N_m = 50$	$x = 17.5$
59*-63	odd	$a = 1, b = 1, N_m = 50$	$x = 17.5$
65-67	odd	$a = 1, b = 1, N_m = 50$	$x = 19$
69-76	both	$a = 1, b = 1, N_m = 50$	$x = 19$
76-81	both	$a = 1, b = 1, N_m = 50$	$x = 20$
81-82	both	$a = 1, b = 1, N_m = 50$	$x = 22 + 2/3$

All starred values are associated with magic numbers. Where the ranges overlap at 18, 39, 76, and 81 the average of the two midpoints is used. At 50 (tin) both midpoints are valid.

The last changes that occur as x shifts to 19 for both odd and even elements may be the result of obtaining proton stability in the lead shell at terbium for odd elements and ytterbium for even ones. Terbium has 15 protons out of 32 in the lead shell, and ytterbium has 20. Stability should occur with 16 protons. Terbium is acceptable, but stability should have come at dysprosium for even elements.

In the following table, the number of protons, the element abbreviation, the midpoint, and the isotopes for the odd-numbered elements from fluorine to thallium are listed. Isotopes suggested by the model that are not stable are bolded. Stable isotopes not suggested by the model that nevertheless exist are in bold italics. The only errors are with lanthanum and tantalum.

Odd-Numbered Element Isotopes

protons		midpoint	isotopes
9	F	18.67	19
11	Na	23	23
13	Al	27.33	27
15	P	31.67	31

17	Cl	36	35,37
19	K	40	39,41
21	Sc	45	45
23	V	50	51 (no 49)
25	Mn	55	55
27	Co	60 (59)	59
29	Cu	65 (64)	63,65
31	Ga	70	69,71
33	As	75	75
35	Br	80	79,81
37	Rb	85	85
39	Y	90-88	89
41	Nb	93	93
43	Tc	98	(neither 97 nor 99)
45	Rh	103	103
47	Ag	108	107,109
49	In	113	113
51	Sb	122	121,123
53	I	127.33	127
55	Cs	132.67	133
57	La	138	**137**,139
59	Pr	141.5	141
61	Pm	146.83	(no 147)
63	Eu	152.17	153 (no 151)
65	Tb	159	159
67	Ho	164.33	165
69	Tm	169.67	169
71	Lu	175	175
73	Ta	180.33	***180***,181
75	Re	185.67	185
77	Ir	192	191,193
79	Au	197.33	197
81	Tl	204	203,205

$^{180m}_{73}Ta$ is expected to be unstable. The *m* indicates that it is meta-stable, though all attempts to show that it decays have failed. $^{137}_{57}La$ is a mistake with the algorithm which is duplicated by the failure to recognize $^{137}_{56}Ba$ as stable.

The next table provides the estimated stable isotopes for even-numbered elements. I will deduct all of the estimated isotopes from the preceding table except for ${}^{180}_{73}Ta$, the only even-numbered stable isotope of an odd element, and ${}^{123}_{52}Te$, the only case where a stable odd element shares a nuclide with an even element. The explanation was given earlier.

Even-Numbered Element Isotopes

range	p		midpoint	isotopes
3	8	O	16.5	16*,17,18
3	10	Ne	20.83	20,21,22
3	12	Mg	25.17	24,25,26
3	14	Si	29.5	28,29,30
5	16	S	33.83	32,33,34,36*
5	18	A	37.83	36,38*,40
7	20	Ca	42.5	40*,42,43,44,46
5	22	Ti	47.5	46,47,48,49,50
5	24	Cr	52.5	50,52,53,54
5	26	Fe	57.5 (56)	54,56,57,58
7	28	Ni	62.5 (61.5)	58,60,61,62,64
7	30	Zn	67.5	64,66,67,68,70
7	32	Ge	72.5	70,72,73,74
7	34	Se	77.5	74,76,77,78,80
7	36	Kr	82.5	80,82,83,84,86*
7	38	Sr	87.5	84,86,87,88*|
7	40	Zr	90.5	|90*,91,92,94
7	42	Mo	95.5	92*,94,95,96,97,98
9	44	Ru	100.5	96,98,99,100,101,102,104
9	46	Pd	105.5	102,104,105,106,108,110
9	48	Cd	110.5	106,108,110,111,112,114
10	50	Sn	115.5-118.83	112,114,115,116,117,118,119,120,122,124
9	52	Te	124.17	120,122,123,124,125,126
9	54	Xe	129.5	126,128,129,130,131,132,134
9	56	Ba	134.83	132,134,135,136,***137***,138*
9	58	Ce	140.17	136,138,140*,142, **144**
9	60	Nd	144.17	|142*,143,145,146,148
9	62	Sm	149.5	144*,149,150,152,154
9	64	Gd	154.83	154,155,156,157,158,160

9	66	Dy	160.17	156,158,160,161,162,163,164
9	68	Er	165.5	162,164,166,167,168,170
9	70	Yb	172.33	168,170,171,172,173,174,176
9	72	Hf	177.67	176,177,178,179,180
9	74	W	183	182,183,184,186
9	76	Os	188.83	187,188,189,190,192
9	78	Pt	194.67	192,194,195,196,198
9	80	Hg	200	196,198,199,200,201,202,204
9	82	Pb	208	204,206,207,208*|

The minimum ratio of protons to neutrons is found in the largest stable isotope of mercury, ${}^{204}_{80}Hg$. If it were possible to continue adding neutrons to lead, there would likely be isotopes of lead that would be stable with 210 and 212 nucleons.

Ranges are truncated to the left due to neutron-magic numbers for zirconium and neodymium, to the right for strontium and lead, but not truncated to the right for cerium or argon because the magic isotope is within $1/6$ of the midpoint.

In addition to the expected error at ${}^{137}_{56}Ba$ because ${}^{137}_{57}La$ had been chosen as stable, there are 2 more errors, ${}^{180}_{73}Ta$ and ${}^{144}_{58}Ce$. Without double-counting for the lanthanum mistake there are 3 errors for 238 stable isotopes. If we then exclude the metastable tantalum, there are only 2 errors for 99.2% accuracy.

Isotope Abundance

It is possible to develop some general rules that help to find those isotopes for an element that represent at least 20 percent of the total abundance, and also determine when there is a dominant isotope with more than 50 percent of the abundance. Those even elements with equal numbers of protons and neutrons up to and including calcium, with the exception of argon, are the most abundant isotopes.

Doubly magic isotopes are dominant. Neutron-magic isotopes are either dominant or are equal to the next most important isotope. Exceptions are ${}^{16}_{36}S$, ${}^{18}_{38}A$, ${}^{42}_{92}Mo$, and ${}_{144}^{62}Sm$. The midpoint is the next most important determinant of abundance, though it only applies to even elements and only for lighter elements. Starting with germanium, the even isotope following the even isotope closest to the midpoint becomes equally or more important. In addition to the rules, there is one particular case of dominance that appears to be associated with nickel 56 being at the end of the fusion chain, $^{56}_{28}Ni$. The following table has numerous errors suggesting that a simple model, even if it is useful in determining stable isotopes, is insufficient at explaining which isotopes are the most abundant.

Only isotopes above 20 percent are listed. When there is a dominant isotope it is listed alone. When no isotope is dominant, two or three of the most important are listed. Those not estimated by the model are in parentheses. If the dominant isotope is the even isotope following the midpoint isotope, it is in bold italics. All isotopes that are n-magic are starred. All dominant isotopes that are not estimated by the model are in bold. There is no attempt to explain the abundances of the odd elements that have two isotopes, whose isotopes are usually either equal, or one isotope is about one third as likely and the other two thirds.

Isotope Abundance

element	isotope(s)	element	isotope(s)
O*	16	Pd	(105),106,108
Ne	20	Cd	***112***,(114)
Mg	24	Sn	***118***,120
Si	28	Te	(128),(130)
S	32	Xe	(129),(131),***132***
A	**40**	Ba*	138
Ca*	40	Ce*	140
Ti	48	Nd*	142
Cr	52	Sm	***152***,(154)
Fe	56	Gd	156,158,(160)
Ni**	58	Dy	***162***,(163),(164)
Zn	**64**	Er	166,(167),168
Ge	72,74	Yb	172,174
Se	78,80	Hf	178,180
Kr	***84***	W	182,184,(186)
Sr*	88	Os	***190***,(192)
Zr*	90	Pt	194,(195),196

Mo	***98***	Hg	200,202
Ru	***102***	Pb*	208

Starred elements indicate that the isotope is neutron-magic. Nickel is double-starred due to its status of ending the fusion sequence. Starting with titanium and excluding all of the starred elements, the midpoint isotope reaches 20 percent in only thirteen out of twenty-four elements. The even isotope following the midpoint isotope, starting from germanium, reaches at least 20 percent in nineteen out of twenty elements. Although the midpoint is successful at estimating which isotopes are stable, the greater abundance of the isotope with 2 more neutrons suggests that, even as early as germanium, the ratio of protons to neutrons corresponding to the midpoint is too high.

18.4 Island of Stability after Lead

The fourth shell is full of neutrons at ${}^{208}_{82}Pb$, but there are no protons in the shell. Adding another neutron would require starting a fifth shell. The shell model described in this chapter predicts the sudden end of stability at ${}^{208}_{82}Pb$ due to the inability to create a fifth shell without protons in the fourth shell. It can also predict a zone of relative stability after lead. The most stable arrangement of protons in a shell is 6, with 4 along the equator and one at each of the poles. Radium will be the beginning of a small island of stability that stretches to curium. There is no simple equation as there was for stable isotopes. It is necessary to use the complete shell model. The shells of interest are the lead and after-lead shells.

	0	2	***3***	4	***6***	9	***12***	14	***12***	9	***6***	4	***3***	2	0	
1	2	***3***	4	***6***	9	***12***	14	***16***	14	***12***	9	***6***	4	***3***	2	***1***

It is necessary to add protons to the top row and neutrons to the bottom row. Protons and neutrons must be spaced as far apart as possible in their respective shells, but the protons in the inner shell must be able to attach to two neutrons in the outer shell. For example, a proton in the 12 column of the top row must connect to a neutron in the 9 column and one in the 14 column or two in the 14 column. The placement of the neutrons is the limiting factor because it is necessary to add two neutrons for each proton and the neutrons cannot be next to each other.

To find the number of neutrons available, we divide the column number by 2 and round down to the next even number. In this way, 6 neutrons can be placed in each of the two 14 columns, 4 in each of the two 9 columns and 2 each in the two 4 columns, which gives 24 neutrons, enough to reach plutonium 244.

The first 6 protons are easy to place with 1 each in the two 3 columns of the upper row and 2 each in the two 12 columns. The 2 protons in each of the two 12 columns pull 2 neutrons from the two 14 columns in the bottom row leaving only 2 more neutrons available in each of the two 14 columns. The 1 proton in the 3 column cannot take both neutrons from the 4 column below because they are on opposite sides of the sphere and must take 1 neutron from the 4 column and 1 from the 2 column. Radium 226 is then done. Adding one proton to each to of the 6 columns and taking 2 neutrons from the 9 columns below them reaches thorium 232. Adding an additional proton to each of the 12 columns where each takes 1 neutron from the 14 column below it and 1 from the 9 column completes uranium 238. Adding a proton to each of the 6 columns and taking 1 neutron from each of the 4 and 9 columns reaches plutonium 244.

At plutonium 244 three of the neutron columns are exhausted in the lower row: 2, 4, and 9. There are two neutrons still available in the 14 columns, but they are too far apart to be used by the same proton. Curium can easily add 2 additional protons around the center, 1 proton in each of the 12 columns reaching the maximum of 6 protons in each of the 12 columns, but it can connect to at most one neutron in the 14 column below. Curium 247 and 248 are then possible, but curium 249 and 250 are much less stable because the added protons are unable to connect to the 2 last neutrons. By determining what locations are available as the after-lead shell grows, it is possible to understand why the island of stability exists, and also why curium is not most stable at 250 unlike all the other elements in the island whose most stable isotope includes 2 neutrons for every added proton starting from radium.

Chapter 19

Fermi's Golden Rule

The Golden Rule has three sections. In the first section, there is a multiplier that includes the incoming masses or mass, an $\hbar^2$ or $\hbar^{-1}$, and a statistical factor to manage multiple copies of the same outgoing particle. The central section is the wave function squared, the most difficult part. The third section is devoted to a series of terms that guarantee the conservation of momentum and energy. The precise way that particles scatter or decay is managed by the wave function—the middle piece. The remaining two pieces make up the phase space. When the incoming mass or masses are substantially greater than the outgoing masses, the phase space is larger. In the case of the Golden Rule for decays, a larger phase space is usually associated with a larger probability for that particular decay channel. The Golden Rule for decays is used to determine the decay rate for a particular decay channel. The Golden Rule for scattering is used to find the scattering cross section associated with the collision of two or more particles.

SUiTe requires the conservation of energy and momentum. Incorporating them into the two Golden Rules seems obvious, but in the case of decays, the Golden Rule suggests that all possible decay channels where the outgoing masses are less than the incoming mass should be possible. As we saw in chapter 11, on kaon branching ratios, the structure of the kaon itself guarantees that the outgoing masses will always be less than the original particle mass. Since all mesons, and the bosons of the weak force, are constructed from pions, all these initial particles will decay into subsequent particles with less total mass than the decaying initial particle. The particles ultimately decay into pions which have less mass than the particle that decayed, and all pions decay into particles with less mass than a pion.

All baryons other than the neutron and proton also include pions, and they will also decay into particles with less total mass. Even for atomic nuclei that spontaneously decay, the incoming mass is always greater than the sum of the outgoing masses. It isn't clear that it is necessary to guarantee that the incoming mass-energy is greater than or equal to the outgoing mass-energy in decays. A correct theory based upon the structure of the incoming particle will do the same. The structure of the incoming particle also limits the possible decay channels, excluding most of the minor decays as noise that are presently counted as valid decays.

19.1 Golden Rule for Decays

The formula is:

$$\Gamma = \frac{S}{2\hbar m_1} \int |A|^2 (2\pi)^4 \delta^4 (p_1 - p_2 \ldots - p_n) \times \prod_{j=2}^{n} 2\pi\delta(p_j^2 - m_j^2 c^2)\theta p_j^0 \frac{d^4 p_j}{(2\pi)^4}$$

Γ is the decay rate, *S* is the statistical factor, *A* is the amplitude (wave function), δ is the Dirac delta function, θ is the Heaviside step function, and *p* is the four-vector of the momentum. p^0 is the zeroth element of the four-vector and represents the mass-energy.

If all possible decays—meaning all decays where the incoming mass is greater than the sum of the outgoing masses—were equally likely, this formula would be more useful. Because the possible decays depend entirely on structure in SUiTe, it is better to determine the decay channel or channels directly from the structure of the incoming particle. Once the decay products and their probabilities are known, it is possiable to determine minimum and maximum momenta for each of the decay products without the need to refer to the above formula.

The amplitude in the Golden Rule is constructed by combining the terms corresponding to the associated Feynman diagrams which show the possible ways in which the particle could decay. For the kaon, where all the 12 major decay channels are through the weak force, the diagrams would include the W and Z bosons. In chapter 11, we discovered that the bosons of the weak force were not involved in calculating the branching ratios. Also in chapter 11, there was something similar to the strong force coupling constant, and it was necessary to take successive powers of that constant, though not in the way perturbation is done in QED. Perhaps the key to a coupling constant greater than one, is to limit the powers of the constant as was done with kaon decays.

For electromagnetism there are factors of the fine structure constant of $\alpha, \alpha^2, \alpha^3 \dots \alpha^n$ used in constructing the wave function. Since α is much less than 1, each additional term adds less and less to the total. The multiplicative factors in chapter 11, chapter 12, and in section 20.3 for the mesons, massive bosons, and the tauon, are greater than 1. For multiplicative factors greater than one, there must be an endpoint. In chapters 11 and 12, the factors—which end at the fifth power—provide all the known major branching ratios of kaons (chapter 11), and the tau and D mesons (chapter 12.) In chapter 20, each successive power helps determine the masses of particles ranging from the kaon to the Higgs boson.

19.2 Golden Rule for Scattering

The formula is:

$$\sigma = \frac{S\hbar^2}{4\sqrt{(p_1 \cdot p_2)^2 - (m_1 m_2 c^2)^2}} \int |A|^2 (2\pi)^4 \delta^4(p_1 + p_2 - p_3 \dots - p_n) \times \prod_{j=3}^{n} 2\pi\delta(p_j^2 - m_j^2 c^2)\theta p_j^0 \frac{d^4 p_j}{(2\pi)^4}$$

This formula is for two incoming particles that could potentially result in more than two outgoing particles. The cross section is σ. Cross sections have units of area. There is no reason this formula shouldn't work, and it does work beautifully. In physics, energy and momentum are both conserved. If an electron comes in, an electron or its energy comes out. Because an electron is not a particle that occupies a specific location of space but is a wave-particle, it is necessary to speak of collisions as involving an area over which particles interact with each other. The interactions are described using Feynman diagrams and make up the amplitude.

The only way in which the Golden Rule for scattering could fail is if we leave something out. Unfortunately, whenever particles annihilate, we do leave something out—namely, dark matter. The opposite is not true because the creation of particles from energy, while our image may be false, nevertheless requires sufficient energy to both break up particles of dark matter and then reassemble them into a new form. The energy accounting works even though we were not aware of the process that had occurred.

The one clear failure of the Golden Rule for scattering is in the modeling of the annihilation of an electron and a positron. The two particles can form positronium, a temporary atom with the properties of hydrogen. The electron and positron spiral toward each other and annihilate. The resultant energy equals their masses plus their momentum. The electrostatic potential energy is missing, even though it is of the same magnitude as the masses. This should have told us that something was wrong. Nature doesn't throw energy away.

The Golden Rule for scattering works well even when additional particles are created as the result of the kinetic energy of the incoming particles. When the particles come in from a distance and return to a distance, it is possible to assume that the electrostatic potential before and after the reaction is the same. This is not the case when the number of charged particles going into the reaction is greater than the number of charged particles leaving the reaction. In that case, the electrostatic potential must be measured. The missing energy must be explained. It is in these processes that it will be necessary to add particles of dark matter to account for the missing energy.

Chapter 20

The Mass and Structure of Particles

The six stable particles of our universe are the electron neutrino, dark matter, the electron, the positron, the proton, and the antiproton. The electron neutrino and dark matter are their own antiparticles. The mass of the electron is due to the interaction of the electron with the universe-wide electrostatic field. The mass of dark matter is likely to be less than that of an electron.

The protino is a building block of the proton and is positively charged. It supplies nearly all of the mass of the proton. There are 5 protinos and 4 electrons in the proton. The mass of the protino depends on the mass of the electron and dimensionless fundamental constants.

$$m_{pr} = 10 \cdot c \cdot \mu_0/\pi \cdot m_e/\pi \cong 195\ MeV$$

c has two forms in SUiTe, one with the usual units and one that is proportional to the linear ratio of the number of the more common particles of space (the d-dual) to the less common particles of space (the w-dual). In this second form, c is dimensionless. Like c, μ_0 has a form that is dimensionless in SUiTe. μ_0 is proportional to the ratio of the force of the d-dual to the force of the w-dual. In meson construction, one-fifth of the value of m_{pr}, or an integer multiple thereof, is used. For the purpose of baryon construction, other than protons and neutrons, I will define a value, $M_{0.4} = 0.4m_{pr} \cong 78\ MeV$. Simple fractions of this value will be used in determining the masses.

The pion is the second smallest unstable particle and the smallest of the mesons. The pion is used to build all mesons larger than a pion and is necessary in the construction of those baryons that are larger than the neutron. When pions are joined to a nucleon, either a proton or a neutron, the resulting particle is a baryon. Any particle that contains a nucleon, with or without additional pions, is a baryon. When pions are used by themselves, the result is a meson or a massive boson. The tau is constructed in the same way as the mesons.

20.1 Baryon Structure and Mass

For unstable particles, the total mass is greater than the mass of its parts. This is not true of the proton (see section 20.5) or of atoms where the mass of the parts is greater than the total mass. It is necessary to deduct the binding mass from the mass of the parts. Stable particles require energy in order to be broken into parts. Unstable particles give off energy when they decay.

All baryons larger than a proton decay. Baryons can be made with a single direct link between a nucleon and another particle, but most are made in the shape of triangles. A Λ (lambda particle) is a nucleon and a pion held together along a line that adds 39 MeV to the mass, or $0.5 \cdot M_{0.4}$. The Ξ and Ω are both in the shape of triangles. Charmed and bottom versions of these three baryons are constructed in the form of triangles.

Four baryons are estimated in their regular, charmed, and bottom versions. The following table contains three of the four. The N in the table below is either a proton or a neutron and will be given a mass of 938.9 MeV. K is a kaon with an average mass of 495.6 MeV. π is a pion assigned a mass of 137.3 MeV. η is an eta meson with a mass of 547.9 MeV. ρ is a rho meson with a mass of 775.3 MeV. D is a D-meson with a mass of 1869.6 MeV, D_S is a strange D-meson with a mass of 1968.3 MeV. K^* is $K^*(892)$ kaon with a mass of 891.7 MeV. The fractions stated are fractions of $M_{0.4}$.

	$\boldsymbol{\Lambda}$			$\boldsymbol{\Xi}$			$\boldsymbol{\Omega^-}$	
	N			N			N	
		$^1/_2$	$^1/_2$		$^1/_2$	$^1/_2$		$^1/_2$
		π	π	$^1/_3$	π	π	$^1/_3$	K
	$\boldsymbol{\Lambda_c^+}$			$\boldsymbol{\Xi_c}$			$\boldsymbol{\Omega_c^0}$	
	N			Λ			Λ	
$^4/_3$		$^4/_3$	$^4/_3$		$^4/_3$	$^4/_3$		$^4/_3$
η	$^4/_3$	K	η	$^4/_3$	K	ρ	$^4/_3$	K

	$\boldsymbol{\Lambda_b^0}$			$\boldsymbol{\Xi_b}$			$\boldsymbol{\Omega_b^-}$	
	Λ_c^+			Λ_c^+			Ω_c^0	
$^5/_2$		$^5/_2$	$^7/_2$		$^7/_2$	$^5/_2$		$^5/_2$
D	$^7/_3$	K^*	D_S	$^4/_3$	K^*	D_S	$^4/_3$	K^*

All of the entries in the above diagrams for particles are added to obtain the mass of the particle. For example, the mass of $\boldsymbol{\Lambda_b^0}$ is:

$$\Lambda_b^0 = \Lambda_c^+ + D + K^* + (^5/_2 + {}^5/_2 + {}^7/_3)M_{0.4}$$

I have not included the Σ baryon in the table because it is very simple. The structure of the Σ baryon is always a line connecting two particles, one of which is always a pion. Both the charmed and bottom Σ baryons are short-lived and decay by dropping the pion. The connection energy to the pion is $^1/_3 M_{0.4}$ in Σ_c and $^2/_3 M_{0.4}$ in Σ_b. The Σ is made by linking an N to a pion with $^3/_2 M_{0.4}$. Unlike Σ_c and Σ_b, which have very short lifetimes, Σ has a mean lifetime consistent with the other baryons in that group.

The following table provides the estimated masses and formulas. In the formulas, N, π, K, and the other symbols refer to the masses, in MeV, of those particles.

Baryon	Estimate	Measured*	Formula
Λ	1115.2	1115.7	$N + 0.5M_{0.4} + \pi$
Σ	1193.2	1193.0	$N + 1.5M_{0.4} + \pi$
Ξ	1317.5	1318.3	$N + (4/3)M_{0.4} + 2\pi$
Ω^-	1675.8	1672.4	$N + (4/3)\,M_{0.4} + \pi + K$
Λ_c^+	2294.4	2286.5	$N + 4M_{0.4} + \eta + K$
Σ_c	2449.8	2453.6	$\Lambda_c^+ + (1/3)M_{0.4} + \pi$
Ξ_c	2471.2	2469.4	$\Lambda + 4M_{0.4} + \eta + K$
Ω_c^0	2698.6	2695.2	$\Lambda + 4M_{0.4} + \rho + K$
Λ_b^0	5619.8	5619.6	$\Lambda_c^+ + (22/3)\,M_{0.4} + D + K^*$
Σ_b	5808.9	5813.4	$\Lambda_b^0 + (2/3)M_{0.4} + \pi$
Ξ_b	5796.5	5793.2	$\Lambda_c^+ + (25/3)\,M_{0.4} + D_S + K^*$
Ω_b^-	6049.2	6046.1	$\Omega_c^0 + (19/3)\,M_{0.4} + D_S + K^*$

*The measured values are taken from the P.A. Zyla et al. (Particle Data Group), Prog. Theor. Phys. 2020, 083C01 and 2021 update. When there are multiple values for a baryon, I have taken simple averages by first taking the average of the charged particles, and then taking the average of the result and the neutral particle. All values have been rounded to a single decimal place.

In the formulas, with the exception of $M_{0.4}$, which is 78 *MeV*, I have used either the sole value or the averaged value rounded to one decimal place from the 2021 *Review of Particle Physics* for the particle used. For *N*, I have taken the average of the proton and the neutron. I have used the smallest value for K^*, the one that results from the charged-hadron decay path. In order to avoid propagating errors, I have used the measured values in the formulas rather than the estimated values from the table.

20.2 Meson Structure

The lightest meson is the pion. Its structure is like the structure of the proton, a three-ribbed sphere. Pions are used to construct both baryons and mesons. For mesons, the energy necessary to make connections between pions can be estimated by using powers of the positive, real, fourth roots of 2 and 3.

The value of the pion mass that is used in building up the mesons is a mix of neutral and charged pions. I will use 138.8 *MeV*. Different mesons favor neutral over charged pions in their structure, and the average pion mass value is then lower. In the following table, I will include the value of the average pion mass that results in the measured mass for that particle.

Pions can match spins with other pions according to the rules laid out in chapter 11 on the branching ratios of the kaon. The kaon is the most stable meson constructed of three pions, but it isn't the only one. The eta meson is also constructed of three pions. When nature constructs more than one particle with the same number of pions, the particle with the lowest mass is the most stable.

As I have already noted when deriving the subshell structure of the electrons of an atom when the atom has been exposed to a magnetic field, and also in the construction of the shells of nucleons in the nuclei of atoms, the square is the ideal form for spin-matching. A structure in the shape of a square, as well as larger squares made of smaller squares, with three sides filled or with all four filled, is the dominant structure used by mesons.

For particles smaller than *D*-mesons, there is enough information to show that nature constructs one or more particles for each number of pions. There are particles with three pions such as the kaon and eta mesons, and with four pions such as the rho and omega mesons.

Squares allow particles to match their spins with both adjacent and opposite neighbors. The alpha particle has a higher than expected binding energy because of the spin-matching between all four nucleons.

20.3 From Kaons to the Higgs Boson

The kaon is stable because of its structure. Three of the sides of the square are filled with pions. A fourth pion on the unfilled side of the square does not appear to be able to easily match its spin with the other three. The four pion particles that attempt to do this, the ρ, ω, η', and $K^*(892)$ mesons, have short lifetimes. The ratio of neutral to charged pions is fixed for the ρ meson at two-to-two. The resulting estimate is much more accurate than using the averaged pion mass value of 138.8 *MeV*.

Meson	Mass	Est.	m_π^*	Formula
$K^\pm$	493.68	495.2	138.38	$(3\sqrt[4]{2})m_\pi$
η	547.86	548.0	138.76	$(3\sqrt[4]{3})m_\pi$
ρ	775.26	776.5	137.05	$(2\sqrt{2})m_\pi^\pm + (2\sqrt{2})m_\pi^0$
ω	782.65	785.2	138.35	$(4\sqrt{2})m_\pi$
$K^*(892)$	891.66	892.1	138.73	$(3\sqrt[4]{6} + \sqrt{3})m_\pi$
η'	957.78	961.6	138.24	$(4\sqrt{3})m_\pi$
$a_0(980)$	980±20	979.6	?	$\sqrt[4]{3}(3\sqrt[4]{3} + \sqrt{2})m_\pi$
$f_0(980)$	990±20	981.5	?	$(5\sqrt{2})m_\pi$
ϕ	1019.46	1023.3	138.27	$(2\sqrt[4]{6} + 3\sqrt{2})m_\pi$
τ	1776.86	1774.25	139.01	$\sqrt{2}\,(9m_\pi + m_e + m_{pr}/40)$

$D^{\pm}$	1869.65	1864.78	139.16	$(9\sqrt{2}+\sqrt{2\cdot}0.5)m_{\pi}$
$D_s^{\pm}$	1968.34	1962.9	139.18	$(9\sqrt{2}+\sqrt{2})m_{\pi}$
D(2010)	2010.2	2007.0	139.02	$(9\sqrt{2}+\sqrt{3})m_{\pi}$
$\eta_c(1S)$	2983.9	2983.9	138.78	$(12\sqrt[4]{2}^{3}+\sqrt[4]{3})m_{\pi}$
J/ψ	3096.90	3100.0	138.66	$\sqrt[4]{3}(12\sqrt{2})m_{\pi}$
$X_{c0}(1P)$	3414.71	3416.1	138.74	$\sqrt[4]{2}(9\sqrt[4]{6}+3\sqrt{3}+\sqrt{2})m_{\pi}$
$B^{+,0}$	5279.33	5294.7	139.24	$\sqrt[4]{2}(21\sqrt{2}+2\sqrt[4]{2})m_{\pi}$
B_s^0	5366.88	5369.0	138.74	$\sqrt[4]{2}(23\sqrt{2})m_{\pi}$
$B_c^{\pm}$	6274.9	6302.7	138.19	$\sqrt[4]{2}(27\sqrt{2})m_{\pi}$
ϒ(1S)	9460.30	9474.1	138.60	$\sqrt{2}(24\sqrt[4]{6}+9\sqrt[4]{2})m_{\pi}$
W	80379	80220	139.09	$\sqrt[4]{2}(486)m_{\pi}$
Z	91187.6	91114	138.91	$\sqrt[4]{2}(552)m_{\pi}$
H	125.10	125.07	138.92	$(378\sqrt{2}+288\sqrt[4]{2}+24)m_{\pi}$

The Higgs mass is in *GeV* rather than *MeV*. m_{π}^{*} is the averaged pion mass that would result in an accurate estimate. The $K^{*}(892)$ has different masses depending on the decay path. I have used the smallest mass, as I did with the baryon estimates. The tau meson has an added electron or positron plus neutrinos. The $D^{\pm}$ meson incorporates half of the pion instead of the whole pion in $D_s^{\pm}$. The mass of m_{pr} can be found in section 20.5.

Meson	Mass	Est.	Formula	Sides
$K^{\pm}$	493.68	495.2	$\sqrt[4]{2}(3\pi)$	3
η	547.86	548.0	$\sqrt[4]{3}(3\pi)$	3
ω	782.65	785.2	$\sqrt{2}(4\pi)$	4
η'	957.78	961.6	$\sqrt{3}(4\pi)$	4
$K^{*}(892)$	891.76	892.1	$\sqrt[4]{3}(K+\sqrt[4]{3}\pi)$	4
$^{*}K_1$		919.0	$\sqrt[4]{2}(K+2\pi)$	3
$a_0(980)$	980±20	979.6	$\sqrt[4]{3}(\eta+\sqrt{2}\pi)$	4
$f_0(980)$	990±20	981.5	$\sqrt[4]{2}(K+2\sqrt[4]{2}\pi)$	3
ϕ	1019.46	1023.4	$\sqrt[4]{2}(K+2\sqrt[4]{3}\pi)$	3
τ	1776.86	1774.3	$3\sqrt[4]{2}K+\sqrt{2}(m_e+m_{pr}/40)$	4
$^{*}\tau_1$		1766.7	$\sqrt[4]{2}(3K)$	3
$D^{\pm,0}$	1869.65	1864.8	$\sqrt{2}(3K+\sqrt[4]{2}\cdot 0.5\pi)$	4
$D_s^{\pm}$	1968.34	1963.0	$\sqrt{2}(3K+\sqrt[4]{2}\pi)$	4
$D(2010)$	2010.26	2007.1	$\sqrt[4]{2}(3K)+\sqrt{3}\pi$	4
$\eta_c(1S)$	2983.9	2984.0	$\sqrt[4]{2}(3\omega)+\sqrt[4]{3}\pi$	4
J/ψ	3096.90	3100.1	$\sqrt[4]{3}(3\omega)$	3
$X_{c0}(1P)$	3414.71	3416.1	$\sqrt[4]{2}(3K^{*}+\sqrt{2}\pi)$	4
$B^{\pm,0}$	5279.5	5294.8	$\sqrt[4]{2}(2\tau_1+K_1)$	3
B_s^0	5368.96	5369.2	$\sqrt[4]{2}(2\tau_1+f_0)$	3

$B_c^{\pm}$	6274.9	6302.9	$\sqrt[4]{2}(3\tau_1)$	3
$\Upsilon(1S)$	9460.30	9474.3	$\sqrt[4]{2}(2(J/\psi) + \tau_1)$	3
$^*T$		18889	$\sqrt[4]{2}(3B^{\pm,0})$	3
$^*T_s^0$		19155	$\sqrt[4]{2}(3B_s^0)$	3
$^*T_c^{\pm}$		22486	$\sqrt[4]{2}(3B_c^{\pm})$	3
W	80379	80222	$\sqrt[4]{2}(3T_c^{\pm})$	3
Z	91188	91117	$\sqrt[4]{2}(4T_s^0)$	4
H_1^0	124.97	125.07	$\sqrt[4]{2}(\sqrt[4]{2}(3T) + T + T)$	3
H_2^0	125.35	125.38	$\sqrt[4]{2}(\sqrt[4]{2}(3T) + T + T_s^0)$	3
H_3^0	125.9	125.70	$\sqrt[4]{2}(\sqrt[4]{2}(3T) + T_s^0 + T_s^0)$	3
H_4^0	125.9	126.19	$\sqrt[4]{2}(\sqrt[4]{2}(3T_s^0) + T + T)$	3

In the above table, I have rewritten the formulas to show that more massive particles are constructed from less massive particles. I have replaced the particle masses with the symbols for the particles. I have used the estimated values for these particles instead of the measured values. The particles with an asterisk to the left have either not been found or, more likely, do not exist with that mass. One interesting result is that the unexpectedly large mass of the η' meson in the standard model now follows directly from the formula for the ω meson when the square root of 2 is replaced with that of 3. The last column in the table indicates the number of the sides of the square that contain a particle other than just a string of neutrinos.

The hypothetical *T* particles are necessary as stepping stones to the *W*, *Z*, and Higgs bosons. In keeping with the naming for the B-mesons, I have given these "T-mesons" normal, strange, and charmed variants. The purpose is to show that the *W*, *Z*, and Higgs bosons are part of a family of particles starting from the kaon that tend to be more long-lived than other particles at the same mass level.

The Amazing Higgs Boson

SUiTe creates four Higgs bosons that have approximately the same lifetime. The ATLAS collaboration at CERN has found a Higgs at 124.97 *GeV*. This matches well with H_1^0 in the above table. The CMS collaboration appears to have focused on H_2^0 at 125.35 *GeV* according to their most recent estimate. The original estimate of the LHC prior to improvements in the detectors and increases in the energy appears to have been between H_3^0 and H_4^0. It is possible that all the teams involved in finding the Higgs were doing an incredible job, but with four particles with highly similar decays within 1.2 GeV of each other when only one particle was expected, the different results are not a failure, but correctly represent experimental science.

This is an ansatz.

I began by looking at the decay channels with branching ratios greater than 0.5%. I then looked for a combination of decay products with the most mass and the most particles. I created a structure for the particle that can match that decay, and then checked that it also matched all the other decays that have branching decay ratios greater than 0.5%.

The next step was to relate that structure to a pattern of squares. Squares of particles can match spins which stabilizes them. The Pauli spin matrices are composed of zeros and the fourth roots of 1. The mass is related to the fourth real positive root of either 2 or 3. If 2, the mass is less, but the lifetime is longer.

The mesons are arranged in levels. The kaon is at the first level, the *D*-mesons and tau are at the second, the *B*-mesons are at the third, the *W* boson is at the fifth, and the Higgs boson is at the sixth. The level can be determined by the number of times that the particle formula includes the fourth root of either 2 or 3.

The second level requires a fourth side with a particle and additional structure. The fourth level does as well. This could be because even levels have a total number of pions, after adding 1, that is not divisible by four. For example, the tau has three kaons of three pions each. Adding 1 results in 10, a number that is not divisible by 4. The charmed *B*-meson has three building-block tau leptons ($^*\tau$) and therefore 27 pions. Adding 1 results in 28: a number divisible by 4. It is not necessary to add additional particles other than the neutrino bridge.

A difficulty with the ansatz is the need to find a value for the mass of an average pion. The neutral and charged pions differ by over 4 *MeV*. When it is possible to identify one or more specific particles that must be in the particle, as it was with the rho, it is possible to improve the estimate of the mass by using the known mass of the particle rather than an averaged value.

Another thing we learn from this model is that there may be many more mesons that have not been identified. It seems that in any formula where there is a fourth root of 2, we can replace it with a fourth root of 3. This remains true for those two powers of the fourth roots as long as we pay attention to which level we are on and make sure that the levels of the powers of the fourth roots between particles that make up a meson do not vary from each other by more than 1. For example, in the sequence ω, η', and $K^*(892)$, all second-level particles, we find that in the formula for $K^*(892)$ the fourth root of 2 is multiplied by the fourth root of 3: two levels. With $f_0\,(980)$ and ϕ, the second level is combined with the first level to create lower-mass versions of a *D*-meson. Finally, those meson formulas that multiply the fourth root of 2 or 3 times a set of values in parentheses are going to be more particle-like—easier to measure and identify—than those mesons where a pion or larger meson has simply been added on.

20.4 CP Violation

CP violation is solved in physics by moving to CPT invariance. The surprise with kaons, *D*-mesons, and *B*-mesons is that they violate CP conservation in their decays. With SUiTe, we have additional knowledge. It may be impossible to detect a particle if it is neutral and decays directly into particles of dark matter or into dark-matter pions—which decay directly into particles of dark matter. With SUiTe, we also know that dark-matter pions are far more common than other pions and can be used in larger structures. This was explained in chapter 12, where tau leptons and D mesons use fanton kaons. Kaons and larger particles constructed entirely of dark-matter pions don't exist, but the fanton kaon has two fanton pions and one neutral pion. The violation of charge-parity may not exist because we may not be counting all of the particles involved in the decay. All kaons start with three pions. CP invariance is not violated among kaons.

In SUiTe, time is dependent on mass. If the mass of a particle changes, time will also change by the same proportion. Charge and the change of coordinates in space are independent not only of each other but of mass and time as well. CP invariance makes sense with respect to electric and magnetic fields because of the way both fields act on a charged particle. The failure of CP invariance and the dependence on time in CPT invariance—something that depends on kinetic mass and therefore, indirectly, on the density of space—does not make sense.

If CP violation were widespread and not related to the kaon and those particles that use kaons in their structure, CP violation might be real. Since CP violation is dependent on structure, and particularly on the square structures of a particular family of mesons as described in this chapter, it is doubtful that CP violation exists. The apparent violation is most likely due to the failure to measure all the decay products.

20.5 Proton, Neutron, Muon, and Pion Mass Estimation

The estimations of the masses of baryons and mesons that I have made have been inexact. The method is a successful ansatz with some indications that it may be connected to a better and more accurate method that is not an ansatz. For particles smaller than kaons, it is possible to provide estimates by creating a structure for the particle. The proton, neutron, muon, and both pions have specific structures that allow their masses to be estimated to within a few hundredths of a percent.

The muon is a ring containing an electron (or positron), 2 electron neutrinos, and a muon neutrino. Half of the ring is an electron and two electron neutrinos. The other half is the muon neutrino. A muon neutrino is made from 3 electron neutrinos in the shape of a T. As long as neutrinos are not bent, they have little mass. Of the five neutrinos in the muon, 4 are bent in order to form the ring.

There does not appear to be an easy path from the muon to either the charged or the neutral pion. The half of the muon with the electron or positron is the starting point for either pion, but the other half of the muon, which is the muon neutrino, is not able to constitute a rib of either pion. When combining neutrinos, it is not possible to simply attach a neutrino to another without matching the two charged particles from which a neutrino is made, the protino and the antiprotino. The negative end of a neutrino must always attach to the positive end of another neutrino. Because the muon neutrino is composed of three neutrinos with the middle one perpendicular to the other two, the ends of the two neutrinos that connect to the center neutrino must have the same charge and the connected end of the center neutrino must have the opposite charge. This is why the muon and tau neutrinos possess unique matter and antimatter forms while the electron neutrino is its own antiparticle.

Each pion must have two neutrinos that stick out like the center neutrino of the muon neutrino. This is necessary in order to form kaons which are made of three pions in SUiTe. The neutral pion has two perpendicular neutrinos on the same rib, but the charged pion has one perpendicular neutrino on each of the two ribs that don't have a charged particle. Five neutrinos joined together including two that are perpendicular is a tau neutrino. Like the muon neutrino in the construction of either the muon or the charged pion, it is readily available for use. It appears that three neutrinos cannot join together without at least one additional perpendicular neutrino.

The charged pion is then a three-ribbed sphere with each semi-circular rib joined to the other two ribs at 120 degree angles. The neutral pion replaces two of the ribs of the charged pion with the half of a muon that contains either a positron or an electron. Two of the ribs of the neutral pion are then identical except that one has an electron and the other has a positron. One of the uncharged ribs of the charged pion consisting of four electron neutrinos is replaced by 5 electron neutrinos in the neutral pion. Three of the neutrinos in this rib are bent and the other two stick out. In the rare cases that this last rib of the neutral pion manages to stay together during decay, it is a tau neutrino. There is less bending of the three neutrinos that are in the uncharged ribs of the pions as compared to the uncharged half of the muon. The bending of the three neutrinos can be determined from the muon. For the uncharged half of the muon to bend around 180 degrees, each connection point (there are three) must be 60 degrees. There are five such points in the uncharged ribs of the pions. 36 degrees divided by 60 degrees is 0.6. The total value of the mass of the rib must be multiplied by 0.6. This does not affect the mass reduction of the four connections where the ribs join. This must be 0.25 or twice the amount between the ribs of the proton.

The proton is also a three-ribbed sphere like the pion. This is perhaps why nature uses protons, neutrons, and pions as nucleons to create larger structures. Two of the ribs have an electron in the center. The two poles of the sphere also have electrons. The remaining five particles are the protino, a particle that is important to SUiTe, but which does not exist in the standard model. In part 3 of the introduction, where I describe how matter comes to exist in SUiTe, I referred to this particle as a POS particle. The protino is a little proton. Five of them are responsible for most of the mass of the proton. There is an exact copy of the protino, but with negative charge, in the structure of the antiproton. This is the antiprotino (the NEG particle of the introduction). The proton has 4 electrons and 5 protinos. The antiproton has 4 positrons and 5 antiprotinos.

Only one rib has a protino at the center of the three ribs of the proton and this one rib provides the attachment point for an electron neutrino and an electron. The neutrino and electron must connect to the interior side of the protino. When an electron neutrino and an electron are added to a proton in this way, it becomes a neutron. The other end of the electron in the neutron is free. If this half-free electron is able to share its electric charge with one or more surrounding protons, the neutron becomes stable. If not, the weak link using a neutrino to join the protino of the proton to an electron when forming a neutron will break. Free neutrons are unstable. More generally, all particles, with the exception of dark matter, are unstable if they include a neutrino.

There are images of all five particles that will be estimated in this section, along with dark matter, the muon and electron neutrinos, and the electron and positron, on the front cover of this book. Note that in the picture of the antineutron, the neutrino is missing and the electron is connected to the wrong side of the protino. The model of the charged pion is also incorrect. I hope to fix these errors in a future version of the book.

The following equations seek to explain as much of the mass of the particles as is possible with the minimum number of variables. Earlier attempts to model the masses of the following five particles indicated that two variables dominated the values. One was the positive mass of the protino. The other was a binding mass. Both were close to possible values in terms of the electron mass and three constants which, in SUiTe, can be without dimensions. The speed of light is proportional to the linear ratio of the most common particle of space, the d-dual, to that of the w-dual. The speed of light squared, divided by the magnetic constant and multiplied by π is essential in determining the ratio between the force of the weaker d-dual to that of the stronger w-dual, a ratio needed for the explanation of gravitation. This is explained in more detail in the next chapter on the equation linking the constants. Both the speed of light and the magnetic constant can be dimensionless in SUiTe.

The letters pr refer to the protino. All mass values are in MeV. Although even unbent neutrinos have mass, their mass is less than 1 eV. I have not included their masses in the following equations. It is by bending a neutrino that it takes on mass.

$$4m_e + 5m_{pr} + 6 \cdot 0.5m_r + 4 \cdot 0.125m_r = m_p$$
$$m_p + m_e + 2 \cdot m_{pr}/40 + 0.8m_r = m_n$$
$$m_e + 4 \cdot 0.2m_{pr} + 1 \cdot m_{pr}/40 + 5m_r = m_\mu$$
$$2m_e + 0.8m_{pr} + 6m_r + 4 \cdot 0.25m_r + 0.6(0.6m_{pr} + 3m_r + m_{pr}/20) = m_{\pi^0}$$
$$m_e + 0.4m_{pr} + 3m_r + 4 \cdot 0.25m_r + 0.6(1.25m_{pr} + 6.08m_r) = m_{\pi^\pm}$$

The mass of the electron is a necessary measured constant of this theory. All other masses depend upon the electron and the values of three fundamental constants. The bent neutrinos in both the muon and the pions are $0.2m_{pr}$ each. The connections between bent neutrinos and the electron are $0.5m_r$, but between bent neutrinos are m_r. These masses are reduced in the uncharged ribs.

The binding energy at the angles of the ribs of the proton is one fourth of the connection between the electron and the protino, one eighth. This follows directly from the rules for the electrostatic interaction. Between an electron and a protino it is $0.5m_r$ which must be divided into two because the connection to the ribs is at 120 degrees ($|\cos 120| = 0.5$). Two protinos must share this energy dividing it again in half. The situation for the pions, where the protinos are replaced by neutrinos results in the binding energies being doubled except for the binding energy between an electron and a neutrino or a protino.

m_e		0.51099895
m_{pr}	$10 \cdot \mu_0/\pi \cdot c \cdot m_e/\pi$	195.05219
m_r	$-\alpha^{-1} \cdot m_e/2\pi$	−11.144865

The last mass is the binding energy. That is why it is negative. Mass values are in *MeV*. c is necessary in determining the linear ratio between the two particles of space and, in that form, is dimensionless. In SUiTe, the rest mass of any particle is equal to the gravitational mass of that particle added to the mass due to its interaction with the universe-wide electrostatic field. The gravitational mass is the attraction of the particles of space for massive particles. Space is made of two distinct particles, the d-duals and the w-duals. The d-dual is far more common than the w-dual. The linear ratio of these two particles is proportional to the speed of light. The ratio of their forces is proportional to the magnetic constant. The magnetic constant, in this use, is dimensionless like the speed of light.

m_{pr} is the mass of the protino and the primary source of normal mass in the universe. m_r is the binding energy necessary to hold particles together. It functions as a form of negative mass.

Proton

$$m_p = 4m_e + 5m_{pr} + 3.5m_r = 938.2979$$

There is no leeway in the equation once the structure of the proton has been determined. The proton consists of 5 protinos and 4 electrons in three ribs. Two of the ribs have an electron in the center, and the other has a protino in the center. The proton is held together primarily with electric charge. Connecting an electron with protinos at each end is m_r so each connection is $0.5m_r$. There are four connections where the ribs join. As explained earlier, each requires $m_r/8$. The multiplier for m_r is then 3.5. There should also be a very small binding mass due to the combining of the spins of the three ribs that should also be deducted, but there is no clear way to determine this value.

The above equation suggests that the mass of the proton is dependent on the mass of the electron and fundamental constants. The mass of the electron depends upon the initial conditions of the universe because its mass, in SUiTe, is dependent on a universe-wide electrostatic field. The strength of the electrostatic field depends on the number of charged particles in comparison to other particles in the universe. The formula for the mass of the proton would have been the same even if the initial conditions had been different. The mass values would have changed and the ratio would have changed, but atoms constructed from electrons, protons, and neutrons would still function. There is nothing magical about the ratio between the mass of the proton and the electron.

Neutron

$$m_n = m_p + m_e + 2 \cdot m_{pr}/40 + 0.8m_r = 939.6456$$

$m_{pr}/40$ is assumed to be the mass gain of connecting to either side of the unbent neutrino that connects the free electron to the protino in the rib of the proton. This value is also used in the muon and in the pions. The above equation minus the proton mass should equal 1.2933321 rather than 1.34772. There may be a missing electric charge interaction between the free electron and the proton that helps hold the neutron together whose binding energy should be deducted or the equation is wrong.

Muon and Pions

These three particles have only a limited connection to the proton, but where possible I have attempted to follow the proton and the neutron. For example, the binding energy between the electron and the protino is $0.5m_r$. The same value is used successfully for the binding energy between the electron and the bent neutrino. The binding mass involved in keeping a neutrino bent is m_r. It is necessary to double the binding mass between the ribs of the pions as compared to the proton.

The following models have limited flexibility. For example, the charged half of the muon is common to both pions. All three must then have the same mass for that part of their structure. In the case of the neutral pion, this value is multiplied by two. It is also necessary to have the same value for the rib connections of the charged pion and the rib connections of the neutral pion both of which must be twice that of the rib connections for the proton. The three neutrinos in the ribs of the pions are less bent than those in the uncharged half of the muon. It is necessary to multiply them by 0.6, the ratio of the two different connection angles. The only adjustable value is the binding mass of the bent neutrinos that attach to each other in the ribs of the charged pion. There appears to be no similar binding mass applied to the muon which suggests that there is a problem.

In addition to the restraints imposed by the models of the proton and neutron, I have deliberately chosen to limit coefficients for m_{pr} and m_r in the models of the muon and the pions so that a change to a coefficient must be no less than one tenth, excluding the coefficient for the unbent neutrino, which remains unchanged from that of the neutron, and the somewhat suspicious additional binding mass added to the charged pion. This requires changes to not be less than 0.9 *MeV* to the total mass while the errors are less than one tenth of that value. This again, in addition to the limitations of the models of the structures of the particles, helps to provide some confidence in the importance of the values of m_{pr} and m_r, both of which are dependent on known constants.

Muon

$$m_\mu = m_e + 0.8m_{pr} + m_{pr}/40 + 5m_r = 105.7047$$

The muon is in the shape of a ring. It is the only particle with this shape, but it is more stable than the pion which is a three-ribbed sphere like the proton. An electron or a positron is attached to neutrinos on both ends. This combination then attaches to a muon neutrino—three electron neutrinos in which the middle one is attached at right angles to the other two.

Each of the four bent neutrinos adds $m_{pr}/5$. They also require a binding mass of m_r in order to stay bent. The free unbent neutrino connected at one end (the neutrino in the neutron is connected at both ends) adds $m_{pr}/40$ as in the neutron, but no additional binding mass. Each neutrino connecting to the electron has a binding mass of $m_r/2$. The free-standing neutrino adds two connections of m_r per connection which is no different than the two bent neutrinos connecting to each other joining the muon neutrino to the other half of the muon. In total, $5m_r$.

Pions

$$m_{\pi^0} = 2m_e + 0.8m_{pr} + 7m_r + 0.6(0.65m_{pr} + 3m_r) = 135.0593$$

The charged half of the muon is doubled in order to construct two of the three ribs in the neutral pion. Construction rules require that there be either no free (perpendicular) neutrinos or two free neutrinos on the third rib. In either case there must be three bent neutrinos on that rib. One similarity between charged and neutral pions is that two of the ribs match each other while the third has a different mass than the other two. The three neutrinos in the uncharged rib are bent less so their contribution to the overall mass of the rib will not be as much per neutrino. If we calculate the angle of a normal rib as three connections that span 180 degrees and the strange rib as five connections that span 180 degrees, everything in the strange rib should be multiplied by the ratio of the smaller angle to the larger or by 0.6. The rib is then $0.6 \cdot \left(3(0.2m_{pr} + m_r)\right)$ plus the mass increase of the two unbent neutrinos. If those two are also multiplied by 0.6, we have the above equation

$$m_{\pi^\pm} = m_e + 0.4m_{pr} + 4m_r + 0.6(1.25m_{pr} + 6.08m_r) = 139.5851$$

Of the five particle masses estimated here, the charged pion is the most difficult. It cannot simply double the uncharged half of the muon for its two uncharged ribs. It must follow the path of the neutral pion by including at least three neutrinos in its two uncharged ribs, but in order to match the charged ends of the particles involved it is necessary to add one free neutrino to each rib. This leaves two neutrinos in each rib connecting end-to-end without a free neutrino in between. They connect at a different angle than the two neutrinos connecting end-to-end in the muon. For this reason, it is possible to imagine an adjustment to the binding energy between them. This is the excuse for adding an additional $0.04m_r$ for each of these neutrino-to-neutrino connections. This adjustment is entirely arbitrary.

$m_p = 938.2979$	$m_p = 938.272081 \pm 6$
$m_n = 939.6456$	$m_n = 939.565413 \pm 6$
$m_n - m_p = 1.3477$	$m_n - m_p = 1.2933321 \pm 5$
$m_\mu = 105.7047$	$m_\mu = 105.6583745 \pm 24$
$m_{\pi^\pm} = 139.5851$	$m_{\pi^\pm} = 139.57039 \pm 18$
$m_{\pi^0} = 135.0593$	$m_{\pi^0} = 134.9768 \pm 5$
$m_{\pi^\pm} - m_{\pi^0} = 4.5258$	$m_{\pi^\pm} - m_{\pi^0} = 4.5936 \pm 5$
$m_p/m_e = 1836.2048$	$m_p/m_e = 1836.15267343 \pm 11$

The mass values on the right are taken from the P.A. Zyla et al. (Particle Data Group), Prog. Theor. Phys. 2020, 083C01 (2020) and 2021 update. The proton electron mass ratio is from 2018 CODATA.

All things considered, the models are able to capture much of the variation in the masses of all five particles. Determining the correct structure of these five particles is important for SUiTe as charged and neutral pions are used to make mesons, just as protons and neutrons are used to construct nuclei. Dark matter is the decay product of a slightly different neutral pion with the same elements, mass, and nearly the same structure as the neutral pion described above. Two different neutral pions are required by SUiTe because there are two ways in which the third rib containing three bent and two free neutrinos can be added to the other two ribs. The neutral pion is easier to make than the charged pion because this structure of five neutrinos is the tau neutrino, which is far easier to add to a particle than putting together a muon neutrino with one additional neutrino in order to form one of two identical ribs of the charged pion.

20.6 Dark Matter

A particle of dark matter is a square constructed of an electron and a positron on two opposite sides with two electron neutrinos separating them. The particle is roughly 1.4 femtometers on a side. The electrostatic forcelines of the electron and the positron are held parallel to each other. They will align in electric or magnetic fields, but each of them will attempt to pull the dark matter particle with an equal force in opposite directions. There is an image depicting a dark matter particle on the front cover.

One possible measure of the mass of dark matter at this time comes from the study of coronal heating using the minimum temperature in the chromosphere (see section 6.1). The dark matter mass drawn from the behavior of the chromosphere is roughly 162 keV. There will then be about 33,000 particles of dark matter for every nucleon in the universe. The small size works well in describing the behavior of dark matter in our solar system and in the galaxy, but if the size were somewhat larger or smaller, dark matter would remain equally effective.

Neutrinos have a specific size in SUiTe with a length equal to $4 \cdot \sqrt{A_{pr}/\pi} \cong 1.2 \cdot 10^{-15}$ meters where A_{pr} is the area of a cylinder which could contain a neutrino or electron (see section 21.1 and 13.4). Given that the neutrino and the electron are not lines but are rounded at the ends, the length of the sides of the dark-matter square will be somewhat longer. This length separating the electron from the positron is consistent with an electrostatic binding energy roughly equal to twice the mass of the electron. For this reason, it is necessary to decrease the mass of a particle of dark matter by deducting that binding energy. In SUiTe, a photon of approximately 1.022 MeV is capable of splitting a particle of dark matter. Splitting a particle of dark matter is similar to ionizing positronium. It is not possible to use the methods from the last section to determine the mass of dark matter due to their relatively large error in comparison to the expected mass of dark matter of about 162 keV.

With four fermions in the structure, dark matter is probably a boson. There exists the possibility in SUiTe that because of the rigid structure, it would be possible for the electron and the positron to combine their spins by placing them parallel to each other. If that is possible, then dark matter would have to be a fermion with twice the intrinsic magnetic moment of an electron.

An Experiment to Find Dark Matter

Charged kaons, most of which decay into muons and muon neutrinos, or electrons and electron neutrinos, also produce many fanton pions (the dark-matter pions of chapter 11) that decay directly into dark matter. To create a dark-matter beam, charged kaons can be separated from the output of a collider or accelerator by using a strong magnetic field, and then passing the charged particles of the correct mass through a hole in a barrier. Following the barrier, the charged kaon is allowed to decay and then passed through a strong electric field that will remove all charged particles. At that point, only neutrinos, photons, and particles of dark matter will remain.

A portion of these remaining particles is then passed through holes in two consecutive lead barriers, creating a dark-matter beam. High-energy photons can be directed against the beam, and the scattering of the photons can then be used to determine the momentum of particles of dark matter. A particle of dark matter should split into an electron and a positron when struck by a photon of sufficient energy. If it is possible to raise the energy of the photons fired at the dark-matter beam such that the total energy of the photon plus the kinetic energy of dark matter exceeds 1.022 MeV, positrons should appear.

If particles of dark matter are fermions, they will be deflected in an inhomogeneous magnetic field. This will allow dark matter to be separated from any photons that had been produced from the decay of the kaons. The background level of neutrinos can be determined by closing the holes in the last two barriers that create the dark matter beam. Particles of dark matter are deflected by matter, so only neutrinos will make it through to the detectors after the holes have been closed.

If this experiment, or a similar experiment, is able to detect and measure the properties of dark matter, and if the properties agree with those stated in this section, I ask that a particle of dark matter be named the fanton, after the French word fantôme.

20.7 Particle Annihilation

Matter-antimatter collisions create energy. Energy can be used to create a matter-antimatter pair. This belief began with the discovery of the positron. The positron was predicted in Dirac's theory. The positron can be thought of as an electron with opposite charge and opposite spin but with the same mass, traveling backward in time. If the positron had not been found, the electron-opposite of Dirac's theory would have been simply the result of the mathematical symmetry of the theory.

In SUiTe, a particle of dark matter is the result of electron-positron annihilation. Matter is conserved, not mass. The problem for SUiTe is that the energy of twice the mass of an electron is able to make an electron and its antimatter twin. When the two combine, the energy given off is equal to twice their mass. The accounting is perfect.

For any antimatter particle that is made, including a positron, under the rules of SUiTe, it is necessary to break at least one particle of dark matter. Breaking a particle of dark matter provides the pieces necessary for the construction of matter and antimatter versions of all particles constructed of neutrinos combined with electrons or positrons. In SUiTe, since all particles, including all mesons and all baryons, but excluding protons and neutrons, require pions, and because pions are constructed of neutrinos with electrons, positrons or both, all temporary particles, except for the neutron, start with the breaking of a dark matter particle. Even the muon, the tau, and the massive force bosons are made from dark matter particles.

The energy needed to construct a meson, or those baryons larger than a neutron, is largely the energy involved in bending neutrinos. Because neutrinos have nearly zero mass to begin with, the energy needed to create the particle is equal to the energy released when the particle is destroyed, and the neutrinos straighten back out. The exceptions to this simple and relatively accurate accounting are dark matter, protons, and neutrons.

In some ways, a dark matter particle is no different than positronium. If there is enough energy, the positron and the electron in dark matter will separate. The energy needed to separate them depends on the distance between them and is equal to the potential energy that was released when the positron and the electron came close to each other. The combination of a photon plus the kinetic energy of a dark matter particle can create a positron-electron pair at less than the usually required amount of 1.022 MeV.

Creating a proton-antiproton pair requires more energy than twice the mass-energy of the proton. Unlike an electron-positron pair, the small difference in energy cannot be easily made up by the motion of the particle with respect to the photon. The required energy necessary to create a proton-antiproton pair from the standard model is about 1877 MeV. In SUiTe, it is necessary to break 5 neutrinos simultaneously. If they are not broken simultaneously, the protinos and antiprotinos will rapidly reform into neutrinos. Once the neutrino is separated, the resulting protino and antiprotino particles each have a mass of half of a neutrino, but a full electric charge. They move at very nearly the speed of light until they rejoin. The only way to have 5 or more neutrinos along with 4 electrons and 4 positrons in the same place at the same time is if they had all been part of 4 dark matter particles. The dark matter particles provide the needed 4 electrons for the proton and 4 positrons for the antiproton.

The energy needed to break 5 neutrinos simultaneously is approximately $5 \cdot 390 = 1950\ MeV$, well above the 1877 MeV deemed necessary. It isn't simply a matter of energy, but it is also a matter of having enough dark matter particles in the same place at the same time. As a stream of ionized nuclei moves through a collider at near relativistic speeds, dark matter particles become trapped in the compressed space in front of the stream. When the stream strikes the target, or strikes a similar ionized stream moving in the opposite direction, the dark matter particles are crushed. If it is possible to create a proton-antiproton pair with less than 1950 MeV and without additional kinetic energy, SUiTe is wrong.

Chapter 21

Necessary Constants and Their Relations

Eight constants form a group in which each value can be determined from the other seven. The values of these eight constants, as well as those other constants that are related to them, are unique to our universe. They can be considered fundamental even though there are formulas using some of these eight constants that can be used to arrive at the values of others.

The interdependent constants are:

1. The reduced Planck constant, ħ.
2. The elementary charge, *e*.
3. The speed of light, *c*.
4. The equatorial area of the spherical protino, A_{pr}.
5. The gravitational constant, *G*.
6. The Coulomb constant, k_C.
7. The fine structure constant, α.
8. The repulsion constant, k_R.

$k_R c$ is the matter-to-matter repulsion force constant with units $m^3 s^{-2}$. It plays a role in preventing the collapse of matter into singularities (black holes) and the failure of dark matter particles to penetrate deeply into either the Sun or the Earth. $k_R = \mu_0/\pi c$. k_R is essential in explaining why solar neutrinos can pass through the Earth, but very high energy neutrinos cannot.

The area of a protino, A_{pr} is also the area of a cylinder twice as long as it is wide that can contain a neutrino or an electron. The protino is introduced in section 20.5. $A_{pr} = m_e/\pi$. It is necessary in describing the Planck force that prevents electrons from falling into nuclei (section 13.4).

The division of m_e, the mass of the electron, by π, changes the unit from kg to m^2, an area. A similar thing occurs with μ_0. The units of NA^{-2} have changed with both charge and mass being replaced by m^2.

21.1 Equation Linking the Interdependent Constants

$$G\left\{{}^{c^3}/_{(k_R)^3}\right\} = k_c \sum_{n=2}^{\infty} \alpha^n \cdot \left\{{}^{ce}/_{A_{pr}\hbar}\right\}$$

The units must be explained. The values in braces are the numbers of particles of space per cubic meter on which the gravitational constant and the Coulomb constant act. They represent a natural unit for the two force constants. It is as if G had been multiplied by kilograms squared, and k_c had been multiplied by charge squared. On the right, the term in braces is not simply particles per cubic meter, but particles per cubic meter multiplied by charge divided by kilograms. It is necessary to have a universal unit of mass-energy that will apply to mass created by the interaction of a charged particle with the universe-wide electrostatic field and mass due to the attraction of space for matter. Both G and k_C, after multiplication by the terms in braces, will have same units of $kg \cdot m^3 s^{-2}$.

An important constraint on this equation is that a change in units for mass (from kilograms to pounds for example) will require the same proportional change to charge. There are reasons to believe that this must be true. Like the rest mass via gravitation, electrostatic charge is a central force and subject to the inverse-square law rule. Both are innate, unmodifiable properties of charged particles. We can also see that any force that acts upon massive particles should have a mass equivalent. Light pressure is the result of a photon interacting with a massive particle. The photon has a Compton mass which is its energy divided by c^2. The electron volt, with a numerical value equal to the elementary charge, also has a mass in which its energy is divided by c^2.

In the attempt to further restrict and rationalize the units of physics starting in the 1950s, it was decided to use the cgs-emu system as the basis for the mksA system that is the standard today. All units of the previous system were multiplied by different powers of ten. The choice to make the numerical value of the charge of the electron equal to the electron volt may seem like the earlier mistake of making the pound both a unit of mass and a unit of force. There are important similarities. Charge can be thought of as a unit of force and the electron volt as a unit of mass. But the use of the pound instead of the kilogram primarily affected gravitation. The association of the elementary charge with the electron volt is essential for electromagnetism. As a result, it also has an effect on the speed of light, which is necessary in all physics. It is not clear that it is possible to disentangle the elementary charge and the electron volt without having to make changes to the two remaining units of length and time. Whatever the case may be, it will be assumed that proportional changes to the units of mass will result in the same proportional changes to the units of the elementary charge such that the numerical values of the elementary charge and the electron volt remain equal.

The units inside the braces on the right-hand-side are the number of w-duals per cubic meter, the particles of space on which electric charge acts. These particles are also the medium of the electromagnetic wave just as air is a medium of an acoustic wave. The units inside the braces on the left are the hybrid particles of space per cubic meter that result in the gravitational rest mass and are the medium for gravitational waves.

The summation is part of a simple geometric series in α, the fine structure constant. The first two terms are not included. The first term, α^0, only concerns electric charges and not particles of space. The second term, α^1, is necessary for describing the interaction between charged particles and the w-duals of space. The remaining terms describe the interactions between the w-duals themselves, the particles of space that are involved in electromagnetism. These powers of the fine structure constant are essential to the Feynman calculus for QED.

We will use the following values for the six known constants from P.A. Zyla et al. (Particle Data Group), Prog. Theor. Phys. 2020, 083C01 (2020) and 2021 update. IN the case of the derived constants, A_{pr} and k_r, The errors should be the same as those for m_e and μ_0 respectively.

1. $\hbar = 1.054571817 \cdot 10^{-34}$ exact
2. $e = 1.602176634 \cdot 10^{-19}$ exact
3. $c = 299792458$ exact
4. $\epsilon_0 = 8.8541878128\ (13) \cdot 10^{-12}$
5. $G = 6.676125 \cdot 10^{-11}$ (from equation)
6. $\alpha = 0.0072973525693\ (11)$
7. $A_{pr} = 2.8996069 \cdot 10^{-31}$
8. $k_R = 1.3342564 \cdot 10^{-15}$

The maximum possible value for A_{pr} in SUiTe is limited by the equation for the maximum radius of the proton: $r_p = 18\sqrt{A_{pr}/\pi} \div 2\pi = 0.877\ fm$. Observationally it is $0.842 \leq r_p \leq 0.877 fm$. The radius that corresponds with the value of A_{pr} is consistent with a radius within this range. The matter-to-matter repulsion constant, $k_R c$ is the subject of chapter 5.

A_{pr} is especially important in SUiTe because a wave of light is unusual as a wave. All waves should have both amplitude and frequency, but with light waves, the amplitude has a very different meaning. It helps us to locate the wave. The wave energy depends only on its frequency. If you are standing in the waves near the beach, you will notice that when the amplitude is at its height, you will be pushed toward to the beach, and when it is at the trough, you will be pulled away from the beach. The higher the wave is, the more energy it has. But a pulse of light of a specific frequency has an energy with no connection to an amplitude. If you are hit by a pulse of light, there is no peak or trough. No matter the phase, the energy remains the same and the push remains the same. All electrons create a pulse of light of the same area, A_{pr}, though the pressure varies according to the frequency which depends on the relativistic kinetic energy lost by the electron. It is this remarkable feature of light which allows us to ignore the amplitude of light and determine its energy only from its frequency.

Although not all light is emitted by the bremsstrahlung of electrons, we measure light in those terms. Light, as energy, is a wave without a wave, a source of energy which does not vary as it passes through space.

21.2 Analysis

The Coulomb constant and the gravitational constant are both inverse-square-law force constants. That the Coulomb constant is 20 magnitudes larger than the gravitational constant is one of the unexplained mysteries of physics. More importantly, gravitation and the Coulomb force are unlimited by distance. They are the only two forces which have this property. They are also the only two forces which can generate rest mass. At slow speeds and in weak gravitational fields, the rest mass and the inertial mass are equal. With the exception of the electron and the positron, the rest mass resulting from interaction with the universe-wide electrostatic field is overwhelmed by the rest mass due to gravitation.

The distance at which the weak force and the electrostatic force are equal is believed to be about 10^{-18} meters, an attometer. In SUiTe, the speed of light is proportional to the number of d-duals divided by the number of w-duals in a meter. If the proportionality constant is one, the number of d-duals in a meter is about $3 \cdot 10^{26}$. This would result in about 10^{79} d-duals and about 10^{54} w-duals in a cubic meter.

Gravity works on the combination of w and d-duals. The result is a hybrid particle that is weaker than a w-dual, but stronger than a d-dual. The weaker the d-dual is in comparison to the w-dual, the less the number of hybrid particles per cubic meter given by the value inside the braces on the left-hand-side of the equation.

The two remaining important constants are not among the constants in the equation: the Boltzmann constant and the Hubble constant. The equation linking the constants either does not change at all or changes very little over time. The Hubble constant is moving asymptotically toward the inverse age of the universe in seconds as the universe expands. The Hubble constant is constant in the sense of space at a given time, but it is not a universal constant. It is a measure of the age of the universe that is dependent on the length of a second at the time the Hubble constant is measured.

The Boltzmann constant is definitional. The average kinetic energy of an ideal gas can be given either as a temperature in degrees Kelvin or in joules as energy. The constant translates between joules and degrees Kelvin.

SUiTe suggests that the fundamental constants should be related because they all ultimately depend on the ratios of the fundamental forces and the ratio of the particles of space in our universe. The value inside the braces on the right hand side gives the number of w-duals per cubic meter in the absence of any compression or contraction of space. This value is used in section 13.4 on the Planck force which prevents the electron from entering the nucleus in hydrogen.

$$\left\{ {}^{ce}/_{A_{pr}\hbar} \right\} = 1.570782 \cdot 10^{54}$$

Matter-to-matter repulsion is necessary by symmetry. If space repels space and space is attracted to matter as well as matter being attracted to space, then matter must repel matter. Repulsion must be one of the fundamental constants. Like the Planck constant, it is necessary to multiply the repulsion constant by the speed of light before using it as a force constant.

Most of the difference in magnitudes between the gravitational constant and the Coulomb constant are due to how they act. The Coulomb constant acts on the w-duals, the particles from which the pulses of light—photons—are constructed. The gravitational constant acts on a hybrid particle that is far more common than the w-dual, but less common than the d-dual. It is this hybrid particle that allows gravitational waves to propagate.

The strange behavior of the gravitational plane wave, expanding and shrinking in horizontally and then doing the same vertically, perpendicular to the first, are the result of the relation between the w-duals and the d-duals in the wave. In order that their forces not interfere with each other, they must be perpendicular to each other and out of phase. Although this behavior matches that of the hypothesized graviton, it is the result of two separate particles acting together as the moving compression of space forces them to rotate and combine in such a way that their separate forces—weak and dense forces—don't interfere with each other.

21.3 Universal and Natural Units

The natural units, or Planck units, are based on the meter (French revolution) and seconds (a strange combination of base 60 from the Babylonians and the use of 12 from the Romans). A meter was one forty-millionth of the circumference of the Earth. The second was one eighty-six-thousand-four-hundredth of the time it takes the Earth to complete a rotation. Even the gram, today based upon Avogadro's number of carbon 12 atoms divided by 12, was based on the weight (on Earth) of a cubic centimeter of water. In this chapter I suggest that the elementary charge is not independent of the gram in the present mksA system of units. How do we go from Earth based units to universal units?

The simple solution is to talk in terms of the number of something. Avogadro's number would be a good choice if it was clearly defined as the number of particles of something per unit volume, perhaps the number of silicon atoms in a pure silicon crystal of a certain size. Length would be better as the number of particles per unit length. The speed of light would be the number of particles per unit length divided by the same number of particles per unit time. This becomes possible in SUiTe because the number of w-duals in empty space per unit length is universal when it is understood that any unit of length is simply a number of w-duals. For example, there are approximately $1.162444 \cdot 10^{18}$ particles in a meter in uncontracted space ($3.543136 \cdot 10^{17}$ particles in a foot). No matter what system of units are chosen elsewhere in the multiverse, there will still be the speed of light, the elementary charge, the Planck constant, and the area of their protino whatever they call it.

The time that it takes a single w-dual to rotate in response to a traveling compression of light, a photon, is a single unit of time. The distance between w-duals is a single unit of length. The charge of an electron is a single unit of charge.

The natural units resulting from these universal units can be based on charge $q_q \cong 1.602 \cdot 10^{-19} C$ or the equivalent electron volt in Joules.

$l_q = (ce/\hbar A_{pr})^{-3} \approx 8.603 \cdot 10^{-19} m.$
$t_q = l_q \div 299792458\, ms^{-1} \approx 2.870 \cdot 10^{-27} s.$
$m_q = \{(c^3/k_R^3) \div (ce/\hbar A_{pr})\} \cdot (eV/c^2) \approx 1.287 \cdot 10^{-20} kg.$

This system of natural units has the advantage that the magnitudes, with the exception of a unit of time, are relatively close in value to each other. The unit length, l_q, is also the distance at which the weak force is no longer masked by space and becomes equal to the electrostatic force. The unit of time, t_q, is the minimum lifetime of any possible particle.

If the universe is quantized, as it appears to be, then all units refer to some number of particles or to their forces. The system of units offered here is based on a unit of charge in the form of an electron volt, and the requirement that a change to a unit of charge must be accompanied by a proportional change to a unit of mass. There should be a separate quantum unit of mass similar to a unit of space. The value of mass given above, $1.288 \cdot 10^{-20} kg$, is much too large for this purpose. One electron volt cannot serve this purpose because the mass of a neutrino must be 2 of these mass units, and the mass of the neutrino is much less than an electron volt. This mass is unlikely to be a simple fraction of an electron volt mass because the electron obtains its mass from the universe-wide electrostatic field while the mass of the neutrino comes directly from its interaction with space.

Appendix

The theory explained in this book requires only one assumption: *The universe (multiverse) started from nothing in such a way that it must be able to return to nothing. Nothing* means that there is no space, matter, energy, time, laws, and mathematics—just nothing. You may be thinking that we can't get something from nothing. Showing how it is possible to get something from nothing is the purpose of this appendix. The second part of the assumption can be stated in this way: *all the fundamental entities from which the multiverse is constructed must be able to combine to form nothing.* We could say that the multiverse, taken as a whole, is self-canceling.

This self-canceling is no more than the requirement that if we start with nothing, we have to have nothing when all the pieces are joined. Furthermore, any entity made under these assumptions must be able to combine with an "anti-entity" without any form of assistance coming from outside of the theory. The reason for this self-canceling is to avoid starting a universe from nothing and having something left over at the end, thus requiring an additional explanation that isn't contained within the theory itself. I want to avoid both outside parameters and outside tinkering. Think of it this way: if there is something permanent left over at the end of this multiverse, then this multiverse started from the end of an earlier multiverse, which started from an even earlier multiverse. Then when was the first multiverse? How did it start?

There is a very important connection between mathematics and our universe that allows us to describe the laws of our universe using mathematics. One can then say that the multiverse, of which our universe is a part, is at the same time nothing and the multiverse. It's a matter of whether we choose to add things up.

Say we have a very long but finite string that goes like this: 1+1-1+1+1-1-1-1-1-1+1-1-1-1-1-1-1-1+1+1+1… requiring that there must be as many 1s added as subtracted. Ultimately, the answer is going to be 0. We can look at the final answer or we can make a subtotal, stopping the chain of numbers at some arbitrary point. The total suggests that the multiverse is nothing, but looking at the subtotals, the multiverse is something. The question is time. If time doesn't exist, then all the 1s have been added up, and the multiverse is nothing. If time exists, then all the 1s haven't been added up, and a multiverse exists. This is the form of time that nature recognizes. Physics time is clock time, which is simply the repetitive motion of matter. It has no direction. Entropy provides a direction, but the direction is due to probability—some reactions are more likely than others.

An explanation for time that I enjoy was originally expressed by the science-fiction writer Ray Cummins and has been stated since by many others: "The only reason we have time is so everything doesn't happen at once." As long as all the pieces of this nothing-multiverse haven't been added up, we have a multiverse to play in. But if we were to imagine that, at the level of the multiverse, time doesn't exist, then all the pieces of the multiverse have already been added up, and we are back to nothing.

I call the theory of our universe (multiverse) the *simplest universe theory* (SUiTe) because it would not be possible to construct a multiverse with fewer initial assumptions. A theory that only requires that the multiverse be constructed from nothing in the way suggested here cannot be made simpler. The piece of the multiverse that is our universe must have been constructed from something because we are here, and at the same time it must have started from nothing. Such a multiverse must be created from pairs such that each pair can combine to equal nothing. Already, the multiverse requires mathematics. It has to have the logic of integer arithmetic. The example I gave above, of a long string of plus 1s and minus 1s, is not just one way of constructing something from nothing, it is the only way. Each entity that we create has to have an anti-entity that cancels it out. This is not matter and antimatter. When matter and antimatter combine, they leave energy and particles of dark matter behind. The electrostatic force is a better model for what is happening.

The mathematics of the electrostatic force allow us to add positive charges to positive or negative charges to negative, but when we add negative to positive we get 0, or the nothing that the multiverse started with. We have to translate this into physical terms. Addition of opposite charges is attraction. Addition of like charges is repulsion. For a multiverse to be self-canceling, each of the particles has to be able to interact with every other particle. This won't work if our particles are points, but could work if they were lines. A space that consists of only one line is one-dimensional, and a multiverse that consists of only a point would be zero-dimensional. Now that we have lines, the multiverse is at least one-dimensional. This is still a mathematical universe and not a real one. These one-dimensional lines do not take up space. This multiverse will not last, because every one of the particles has to be on the same line at the same moment since time does not yet exist. They will cancel each other.

We will start with one force acting along a line, the electrostatic force, which will create a multiverse of one dimension. Now we will add a second force and call it the *weak force*. The name will take on meaning later, but for the moment there is no reason to consider this second force as being any weaker than the first, the electrostatic force. If we join the two forces into a particle and then populate the multiverse with these particles, the multiverse will have two dimensions. Self-canceling requires that these particles interact with each other. How will they interact? According to the cosine of their angle of interaction. Why?

Let's first consider the case where two forcelines of the same type interact with each other. If you and a friend were to stand on ice holding perfectly rigid, straight sticks, and tried to push each other away, requiring that the sticks touch each other only at their ends, then you would find that placing the two sticks in a straight line (end-to-end) would give the maximum push. If they were at an angle, the amount of the force that your friend would feel, as you push her away from you, would be proportional to the cosine of the angle at which the two sticks meet. If the forcelines of our universe are similar to rigid sticks, and if they repel each other, then we will reach the same result. There is nothing special about using the cosine. It is simply the only way in which two rigid lines of equal force that repel each other can interact in a plane.

If we want to use two forces, electrostatic and weak, and join them into a single particle, then our two forces will need to join at 90 degrees. We have to choose 90 degrees because that is the angle where the two lines don't disturb each other. The cosine of the interaction is 0 at 90 degrees. At any other angle, they would be either repelled or attracted by each other.

The multiverse must be self-canceling, by which I mean that for every particle that is created, an antiparticle is also created. If the two meet, they disappear. This requires that there be a means by which the two particles can meet. In other words, particles and antiparticles have to attract each other. This is fine for a particle constructed of a single force, but what if particles are constructed of more than one force?

Imagine that different forces—the weak force and the electrostatic force in this example—don't interact with each other. The two forcelines involved in a single particle can then be at any angle, rather than only at 90 degrees where they won't interact ($\cos 90 = 0$). Let's say that I have an electrostatic-positive forceline joined to a weak-positive forceline at 45 degrees. It attracts a particle made from an electrostatic-negative forceline and a weak-negative forceline also at 45 degrees—that is, the antiparticle of the first particle. The electrostatic-positive and the electrostatic-negative forcelines of the two particles combine, but the weak-positive and the weak-negative meet at 90 degrees. I now have particle and antiparticle joining, with electrostatic-positive and weak-positive forcelines joining with electrostatic-negative and weak-negative forcelines, but they don't disappear because the weak-force forcelines of the two particles are 90 degrees apart. Those two forcelines don't interact at all. The self-canceling rule is strong. The electrostatic forceline must act not only with other electrostatic forcelines, but also with weak-force forcelines, forcing the two different forcelines to be at right angles with each other within the particle.

We need some way of making the forces different. But how can we make them different if their only characteristic is that they come in both positive and negative forms? If all forces interact with each other, attracting or repelling in the plane of their interaction, how can the weak force and the electrostatic force be different from each other if their interaction is also in the plane? They can't. The electrostatic force and the weak force must push or pull on each other out of the plane of their interaction. We need three dimensions to deal with particles made of two forces.

We can make up a language, so that we can talk about these fundamental particles. The positive and negative of the electrostatic force can be + and −. The positive and negative of the weak force can be p for positive and n for negative. A particle of +p, positive for both the electrostatic force and the weak force, meets a particle of –n, negative for both forces, and the two particles disappear. Can we do anything with this two-force, three-dimensional universe? No. The two forceline particles, looking a little like plus signs, will find their opposites, join with them, and disappear. Will it help to add a third force?

With another force, I will need another symbol for our fundamental particles. Let's use a (addition) and s (subtraction) for the new force, which I will call the *dense force*, a name which will make sense later. Here's what I am trying to get across by using +, positive, and addition for the three forces: they are not different. All have attraction and repulsion in the plane when interacting with a like force—electrostatic with electrostatic, weak with weak, and dense with dense. All push or pull out of the plane of interaction with an unlike force—electrostatic with weak, electrostatic with dense, etc. When the three forces combine to construct a fundamental particle in what is now a three-force three-dimensional multiverse, the forces are lines and join at right angles. Our three-force particles will look like three-dimensional Cartesian axes.

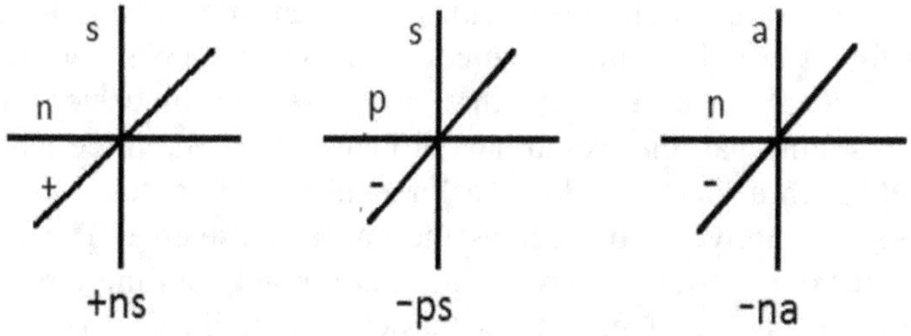

If all the forces are identical, then why is the electrostatic force so strong compared to the weak force? Because there are far more unmasked weak-force forcelines in our particular universe (our piece of the multiverse) than electrostatic forcelines. Things that are rarer are more valuable. Forces that are rarer are more powerful. They are more powerful because they are not screened out by multiple copies of the same force. If I introduce a single positive particle, a proton, into a room full of negative particles, electrons, which one of the electrons will the proton join with? If the same proton comes into a room filled with particles exhibiting the weak force but no electrically charged particles except for one electron, then no matter how distant, the electron and proton will attract each other. We live in a world where the electrostatic force is rare and therefore powerful.

With three forces, there can be eight particles. Why eight? Each force has a positive and a negative. There are three forces, each of which can take on two forms. Each fundamental particle must have all three forces. That gives us $2^3 = 8$ possible particles.

The eight possible particles are, using the symbols I have chosen, starting with the electrostatic force first and dense force last: −ns, +pa, −na, +ps, −ps, +na, −pa, +ns. When a particle meets an antiparticle, the two can successfully rotate and cancel because they can attract and repel in the plane of two dimensions in which like forces meet, but they can also push and pull out of the two-dimensional plane when unlike forces meet. This is not true for four forcelines in four dimensions. With four forcelines, there are rotations possible that will never lead to a particle and its antiparticle canceling. In four dimensions, we could get a permanent something from nothing.

We can't have a multiverse with more than three dimensions if our forces are along lines. We could replace our lines with two-dimensional planes and have three forces in four dimensions, but the resulting particle would have a locus where the three planes meet. Such a locus would be a line and would stretch not only across our universe but across the entire multiverse. The only way for our particle to have a locus at a point, and therefore be limited in its location, is to construct it from lines. You might think that particles with an extended locus that stretched across the multiverse would be interesting, but they would generate contradictions. If such particles existed, we would know about it—or, what is far more likely, we wouldn't exist. Fortunately such particles don't exist. We have reached our first prediction. All of the universes of the multiverse will have to have exactly three dimensions, not more, not less.

This is an important prediction. Three dimensions are obviously what we see, the result of our eyes being formed by millions of years of evolution. If there were more dimensions, we would almost certainly have developed some means of sensing them. What's the experimental evidence? From the smallest we will ever be able to measure to the largest we will ever be able to measure our universe has exactly three dimensions.

We now have eight particles in three dimensions, but we don't have space, energy, matter, or time. On the other hand, we have integer arithmetic, the basis of mathematics. It's time for our first problem. How, when, and where do these eight particles arise?

Let's first consider the question of *where*—space. None of these three dimensional particles take up any space. There is no way to talk about a location, which is always relative to a different location, nor can we talk about when, because physics time and distance are correlated. Physics time requires the motion of matter, but motion requires space. Even the question of *when*, if it could be disassociated from space, could only be a beginning or an end to something. So what is the beginning of space? It occurs when two fundamental particles join with two of their three forcelines. This combination prevents either of the two from joining with their fundamental antiparticles immediately.

We see that the creation of space, when two particles are attracted but cannot completely join, is also the beginning of time. Prior to space, particles and antiparticles could join instantaneously. There was no question of time. Whenever we try to explain space, we are also explaining spacetime.

Now we can answer the questions. The particles can only arise in the absence of spacetime. They will arise the moment spacetime disappears. The creation of these particles ends with the beginning of spacetime. Fundamental particles and antiparticles, because they take up no space and because they attract each other, join instantaneously. The creation of the particles of space, when two fundamental particles attract with two of their forces but repel with the third, also occurs instantaneously. With the creation of spacetime, only those fundamental particles that have been trapped and will eventually become the seeds of particles of matter have survived. All others have been incorporated into space or have combined with their antiparticles and have disappeared. Fundamental particles can only be created in the absence of space. Creation begins and ends at the same instant. The multiverse is born at this same moment.

Our universe comes into existence later. It is necessary for the multiverse to expand to the point that protons and antiprotons become stable. The particles of space repel each other in all the universes of the multiverse. All universes undergo expansion. The moment at which our universe was born was when protons and antiprotons caused our space to separate from a compatible space which was the birthplace of electrons, positrons, protinos, and antiprotinos. In this appendix, protinos are called "POS" particles and antiprotinos are called "NEG" particles. This is necessary because, in the beginning, our universe was combined with a similar universe. At that time, there were also "neg" and "pos" particles.

The expansion of our universe appears to be the same in all directions, suggesting that our universe is probably within a larger universe constructed of a compatible space. It is not possible to state where we are in reference to the rest of the multiverse. At best, we can determine our nearest neighbor universes, but not their nearest neighbors.

How the multiverse was created is more a question of philosophy than physics. The original particles arise in response to the state of nothing because, in that state, there is nothing to prevent them from arising. Perhaps because the arising of the particles from nothing is possible it therefore becomes necessary. This state of nothing is only able to create four particles and their antiparticles. For creation to proceed in any other manner would require additional assumptions. There are 24 different types of space. Each requires two types of dual particles. Even two universes with the same space will not have the two duals of space in exactly the same proportions. No two universes will be alike.

Now we have a question which arises frequently in modern physics: are there infinite universes in the multiverse? I don't want to ruin a seemingly successful branch of theoretical physics, but the answer is clearly no. Particles can only arise in the absence of time and space. If space had not been created, all particles would have instantly disappeared. With the creation of space comes time, and since none of the particles that were responsible for creating space had been created within space and time, we should not imagine that new particles can be created once space and time exist. If that were possible, we would be able to find a location in our universe where space and matter were being created. No such location has been found.

How long did it take to create the space of the multiverse? As stated previously, space was created instantaneously because the combination of particles, both particles with antiparticles as well as those particles that created the duals of space, occurred instantaneously. Although this is difficult to imagine, the initial three-force particles take up no space at all. The three forcelines of the initial particles have only length but neither width nor breadth. Even if there were infinitely many fundamental particles, they would still be unable to create space, yet two fundamental particles when combined with two of their three forces can create a particle of space. The reason that space can exist is due to the repulsion of the two forcelines that repel each other in the dual particle. Their failure to overlay each other creates a tiny space between them. It is not possible to pack dual particles infinitely close. They will always take up space.

Creating Our Universe

Imagine a region in the newly created space where two of the eight particles are far more common than the rest. Let's choose +ns (the positive of the electrostatic force, negative of the weak force, and negative of the dense force) and +pa. We choose the second most common particles to be –ps and –na. The +ns can join with the –ps to create a +ns/–ps. Something similar happens to the +pa and the –na. The two fundamental particles stick together because they both have two attracting forcelines but only one repelling one. The four opposite (attracting) forcelines of the two particles cancel each other, while the remaining two dense-force forcelines—the net "s" forcelines in the +ns/–ps or net "a" forcelines in the +pa/–na—repel each other.

Since we have made the +ns and +pa particles the most common, those fundamental particles that weren't tied up in +ns/−ps and +pa/−na particles can form the slightly less attractive +ns/−na and +pa/−ps particles. We have now created four duals—namely, +ns/−ps, +pa/−na, +ns/−na, and +pa/−ps—using four of the eight particles that are available in our universe. The four particles we haven't used are antiparticles to the ones that we have used. The antiparticles would be more attractive, but once the dual particles of space have been made, any remaining fundamental particles not bound up in particles of matter will have joined and disappeared.

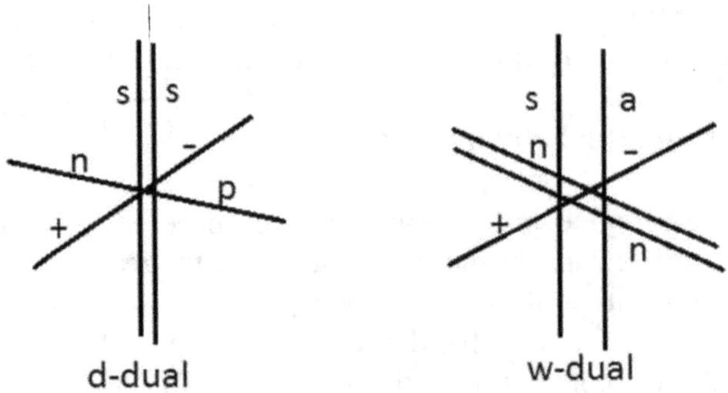

W-duals are responsible for electromagneticism

Let's now consider just two of these dual particles, +ns/−ps and +ns/−na. We could add –ns to the mix. It would be attractive to both of these two duals, but it wouldn't have a strong enough attraction to break up either of these duals. Perhaps it could be stuck in a cage of +ns/−ps and +ns/−na duals. For that matter, if a +ns were still free amid the duals +ns/−ps and +ns/−na, it could also get stuck inside a similar cage.

Does nature do this? At a much grander level the answer is yes. We are alive because hemoglobin traps an iron molecule in a cage. Even from inside its cage, it can attract an oxygen molecule (O_2). If it wasn't inside a cage, the iron and the oxygen would combine to form iron oxide, rust. Because it is inside a cage, it can form weak bonds that allow us to attract oxygen and then move it to where we need it. The weak attraction allows our cells to separate the oxygen from the iron molecule trapped in its hemoglobin cage.

Let's call a trapped –ns in its cage a "neg" particle, and a trapped +ns a "pos" particle. We can do exactly the same for the +pa/−na, and +pa/−ps duals, using –pa and +pa as the particles in the cages. We can call them "NEG" and "POS." These names aren't very creative, but if we think of them in terms of the electrostatic force, they are useful.

We are now using a total of six of our eight particles: +ns, +pa, −ps, −na, −ns, and −pa. The first four don't cancel each other because none are particle and antiparticle, but the last two, which are responsible for the neg and NEG particles, respectively, need to be stuck inside cages or they would disappear.

Initially, we were dealing with fundamental particles that would have canceled each other. Now we are dealing with more complex particles that can trap fundamental particles inside cages that are made from fundamental particles. The trapped particles can attract the trapped particles in other cages to form more interesting particles. For example, a neg particle can attract a pos particle using the electrostatic force of the two caged particles. But here we have a problem. The duals that construct the cage of a neg particle are all net "s" and net "n." The same is true for a pos particle. These forces will repel each other and prevent a neg/pos particle from being constructed.

Let's try again. How about a neg and a POS? Here the attraction is too much. POS particles are wrapped with +pa/−na and +pa/−ps, while neg particles are wrapped in +ns/−ps and +ns/−na. The +pa/−na duals are net "a" and +ns/−ps are net "s." They can combine to form a quad particle, two duals stuck together. But if the particles that make up the cage combine to form quads, then the trapped particle of the POS, +pa, will be free. It will then be able to combine with the previously trapped particle of the neg, −ns. That's a fundamental particle and its antiparticle. They combine and disappear. We can't stick a neg with a POS, either.

What if two neg particles attach on either side of a POS? As they get close to each other, munching away on the POS particle, they will eventually stop because the neg particles repel each other, leaving a sort of half-eaten neg/POS/neg. There would also be a pos/NEG/pos made in the same way. Not to mention the POS/neg/POS and NEG/pos/NEG. The first two of these semi-devoured particles are the electron and the positron of this theory. We can call the other two the ELECTRON and POSITRON. We know what the electron and positron are, but we've never seen the ELECTRON or POSITRON.

There is something distinctly wrong about this theoretical universe we've just made. But before we try to find a way to fix it, let me go over this mind-numbing list of what I call fundamental particles and their combinations. We have been using logic to combine entities that I call forcelines into bigger things that I call fundamental particles. I then again use logic to combine fundamental particles into things I call duals. Confusing the issue even more, when I have managed to combine more than four fundamental particles into something much bigger, I am back to calling it a particle, though not a *fundamental* particle.

With striking imagination, I name the four possible complex particles that can be created neg, pos, NEG, and POS, due to whether or not their essential (caged) particle is electrically negative or positive. I finally combine these complex particles in order to create something we have some familiarity with, electrons and positrons. What do I want you to remember? The four compound particles and their simple-minded names (neg, etc.) because they keep coming up. I will also need you to remember the names of the duals of our space shown in the previous figure and discussed in more detail later in this section.

The four duals, +ns/−ps, +pa/−na, +ns/−na, and +pa/−ps, make up our original space, the space where electrons, positrons, ELECTRONS, and POSITRONS were born. Here things get tricky. When I say that space is made of duals, I don't mean that space is an empty something in which we can find the duals; I mean that space is constructed of particles that are made of forces. An electron is also constructed of forces. It is both a particle and a piece of space.

What is between the +ns/−na duals of our own space? The +ns/−ps duals. And what is between those duals? Absolutely nothing. One way of thinking about this, which is not correct, but works for me, is to consider the +ns/−ps duals as touching. It's not correct because a locus of forces, even though it gives us a useful location for a particle, is a simplification. The forces extend beyond the locus, and they influence other particles through their extension.

Let's call the more common +ns/−ps duals dense-force duals, or *d-duals*, and call the less common +ns/−na duals weak-force duals, or *w-duals*. That simplifies things a little, but our universe still contains four types of duals, and we don't want that many.

The proton is the key to getting rid of the two extra duals, but let me act as though we have already dumped them and continue with the explanation of the space of our universe as it exists now.

The d-duals repel each other with the dense force, since they are net “s,” the negative of the dense force. They will form a three-dimensional lattice similar to what one might find in an ideal crystal. The w-duals will do the same using the weak force, since the w-duals are all net “n,” the negative of the weak force. Since the particles of the lattice made of d-dual particles will repel each other, and since the particles of the lattice made of w-dual particles will also repel each other, and since space is constructed of these two intertwined lattices, what will space do? It will expand.

By choosing which of our eight fundamental particles is the most common for our universe, and then the next most common, we have deliberately created a space where the electrostatic force is masked in both of the duals that make up space. This means that our universe has very few unmasked electrostatic forcelines. When there are a lot of unmasked forcelines in a volume, they screen each other. The screened force appears weaker. This lack of screening of the electrostatic force by space makes the electrostatic force appear to be very strong in comparison to the weak and the dense forces.

Because there are fewer w-duals than d-duals, the weak force is stronger than the dense force. The rarer the force, the stronger it is. Other than their strength due to scarcity, and the screening caused by having many of the same kind of particle in close proximity, all three forces of this model are the same.

There is nothing weaker than the dense force. The distance between the two “s” forcelines in the d-dual is the smallest distance there is. It would not be possible to stick anything into that space. The two parallel “s” lines of the d-dual will act as one. That won’t be the case with the w-dual. The two parallel weak-force forcelines—the “n” lines—will be wider apart. Having the two fundamental particles that make up the w-dual farther apart allows the d-dual to interact separately with each of the “s” and “a” dense forcelines in the w-dual. (See the previous figure showing w-duals and d-duals.)

The “s” and “a” dense-force forcelines in the w-dual are not able to cancel each other because of the repulsion of the two “n” forcelines in the w-dual, which push the two fundamental particles, +ns and –na, that make up the w-dual, farther apart than is the case with the two “s” forcelines pushing apart the +ns and –ps in the d-dual. With the “s” and “a” forcelines of the w-dual spread apart, the “s” forceline within the w-dual is repelled by the far more common d-duals surrounding them, whereas the “a” forceline is attracted. The w-dual will be forced to spin. All the w-duals of our space are spinning. Because the ratio of d-duals to w-duals does not change, either over time or due to forces like gravitation, the rate of spin is the same for all w-duals everywhere in our universe and throughout time.

There is no preferred direction of spin. Clockwise and counterclockwise are equal. If w-duals are responding to d-duals, what does that tell us? When w-duals need to move, the d-duals are forced to move as well. The more common d-duals form the space in which the w-duals function. The w-duals have the weak force, but the “spin” of the electron is also the weak force. The acceleration of an electron compresses the w-duals in front of it. This compression requires the w-duals to pivot so that they are parallel with each other and perpendicular to the direction of motion. When the electron suddenly decelerates, the bundle of w-duals is freed from the electron and becomes the moving pulse of compressed w-duals that we call *light*. As the pulse moves through space, the w-duals in front of the pulse are forced to pivot while those behind the pulse decompress. As the w-duals pivot, the forcelines of the w-duals cut through the forcelines of the d-duals. The number of d-duals that they have to cut through will determine the speed with which they can pivot. In this way, it is the lattice of d-duals that determines the speed with which a compression of w-duals can move. More simply, the speed of light is determined by the ratio of d-duals to w-duals. Other properties also depend on this ratio. Because this ratio does not change, even with the expansion of space, we can predict that the speed of light will not change.

We were able to create electrons and positrons. They would have to be exactly the same except for their electric charge. The only differences between them were the fundamental particles trapped in their cages, and the only difference between the fundamental particles in those cages was the electric charge. In physics, the idea that there may be two particles that are identical except for electric charge didn't appear until 1928, with the Dirac theory of the electron.

To reach this point, we skipped over the early universe, when our universe of electrons and positrons was joined with a universe where there were also particles where every pos particle was replaced by a POS particle and every NEG particle by a neg particle. Our electrons were similar to their ELECTRONS, and our positrons were similar to their POSITRONS. This is not matter and antimatter, where we replace neg with pos and NEG with POS. In our initial combined universe, we can also replace neg with NEG and pos with POS.

Can we make anything bigger than electrons and positrons? Free POS particles and the neg particles at the end of electrons would be a poor combination since they would destroy each other, but if the POS particle could be coated with w-duals and d-duals, both of which are attracted to a POS particle, then the POS particle could safely join to an electron using the electrostatic force.

Let's follow the usual convention about particle symbols and call our electron an e^- and the positron an e^+. Can I make a ring of e^-/POS/e^-/POS/e^-/POS? We're working with forcelines, so this would indicate that the particles join using the electrostatic force at a 120-degree angle. A quick check with the cosine of 120 degrees shows us that half of the electrostatic force is available to hold this ring together. Not bad. But it doesn't work. We haven't talked about *spin* yet, which is the weak force in this theory, but it won't allow this particle to be stable. We can add another half-ring turning this into a three-ribbed sphere. Once again we are using 120-degree angles, but now we are taking advantage of all three dimensions. Making the additional half-ring with two more POS particles and one more electron between them, we find that we have constructed a proton. The ribs can combine their spins (weak force) in such a way as to stabilize this particle. This proton will be net "p" and net "a," the positive of the weak force and positive of the dense force, attracting the w-duals and d-duals of space which are net "n" and net "s."

Until the proton can be constructed, we don't yet have our own universe. Once constructed, protons and antiprotons—a particle identical to the proton except that we switch positrons for electrons and NEG particles for POS particles—will attract our two duals, the w-dual and the d-dual. At the same time there are going to be PROTONS and ANTIPROTONS that will appear and attract their own version of the w-dual and the d-dual, the +pa/−na and +pa/−ps. Our w-duals and d-duals would repel PROTONS and ANTIPROTONS, while their version of our w-duals and d-duals would repel our protons and antiprotons. The previously united space would separate into two distinct spaces with different particles. Both of these two new universes would follow the same laws, though basic constants such as the speed of light will be different because the ratio of their d-duals to their w-duals will be different. Suns and planets would form in both. Life would ultimately appear in both.

If our universe was constructed in this way, there would be two very important temperatures in its creation: the temperature at which protons and antiprotons as well as PROTONS and ANTIPROTONS could be formed, and the temperature at which positrons disappear. When protons and antiprotons can form, the two companion universes separate. Our space will be attracted to our protons and antiprotons, the other space to theirs. A recap:

The universe was created from nothing. The only way to do this was with forces acting along lines. The number of dimensions required was exactly three, no more, no less. Each fundamental particle is constructed of three different forcelines at right angles to each other. Only eight such particles are possible. Four of them are antiparticles to the other four, and when they meet, they disappear. The original universe has broken into smaller universes, which is why it is necessary to refer to the overall universe as the multiverse.

According to SUiTe, our own universe, when still combined with its companion universe, required only six of the eight possible fundamental particles. Electrons, positrons, ELECTRONS, POSITRONS, POS, NEG, pos, and neg particles were created using only these six fundamental particles. Once this early combined universe had sufficiently expanded to the point that protons, antiprotons, PROTONS, and ANTIPROTONS could be formed, the combined universe split into two. The space of our present universe is made of two duals, the w-dual and the d-dual.

We don't care about our previous companion universe except that we hope it was larger than ours and is presently protecting us from some other universe whose space may be made from antiparticles to our space. The recently discovered dark flow could potentially be an interaction with a different universe whose space has antiparticles to ours.

We want to go back to the time when electrons were still swimming about in the four duals of what would become two different universes. The NEG and POS particles are attracting the w-duals and d-duals of what will become our space and attaching them to their outside. With a coating of these two duals, they won't repel each other anymore. They can now attract each other with the strongest force, the electrostatic force. This is how we arrive at the electron neutrino.

We don't think of neutrinos as being sticky, but they are in this theory, using the weak force. The neutrino sticks to one end of the electron and to the POS particle in the middle of the rib of the proton in order to form a neutron. Neutrinos also play a role in the construction of quasi-particles for use in unstable nuclei made from pions that we call mesons. All unstable particles have neutrinos in their structure. Only one particle with neutrinos is stable, the dark matter particle.

In the text, I have introduced eight forces as being necessary for SUiTe: electrostatic, magnetic, weak, dense, matter-to-matter repulsion, space-to-space repulsion (Hubble force), the Planck force, and gravitation. Of these, the electrostatic, weak, and dense forces are the only pure forces. The magnetic force results from the physical manipulation of w-duals, the carrier of the weak force in space. Both matter-to-matter repulsion and space-to-space repulsion are the combined effect of weak-force and dense-force repulsion. The Planck force is due to the compression of space. Gravity is a combined effect of the weak and dense forces because space is attracted to matter and gravitation results from the contraction of space surrounding matter.

The strength of the electrostatic force depends on its rarity, but in the nucleus of an atom, sources of the electrostatic force are packed together. We have nine charges—four negative from the electrons and five positive from the POS particles—in a sphere less than 2 femtometers across. But when we attempt to break apart a proton, the individual charged pieces move farther apart, strengthening the electrostatic force because there are then fewer electrostatic forcelines in a given volume of space. Given what we now know about forces, we should expect the electrostatic force to be weaker in the confines of the proton and to gain strength as the proton expands. The rarer the force, the more powerful it is. This provides the rubber-band effect that is seen in experiments that attempt to tear apart protons.

We have made five stable particles: the electron, the proton, their antiparticles, and the neutrino, which in SUiTe is its own antiparticle. There is also a sixth stable particle in SUiTe but no more after that. All remaining particles will be temporary. They will decay into the six stable particles. The sixth particle is dark matter, the particle I was originally looking for when I decided to try to make a universe from nothing.

A particle of dark matter is wonderfully simple. It is a square primarily held together using the electrostatic force with some help from the weak force. The weak force is the same force that is necessary for the stability of the proton and for the arrangement of electrons in the spherical shells of atoms. A particle of dark matter has an electron opposite a positron, where both have attached themselves to a neutrino at either end. Where e^+ is a positron, e^- is an electron, and ν is a neutrino, we have $\nu/e^-/\nu/e^+$. To see the connections more clearly in terms of their elemental particles, we have:

NEG/POS/neg/POS/neg/POS/NEG/pos/NEG/pos

The energy needed to break a particle of dark matter is the energy needed to pull the positron and the electron away from each other against their electrostatic attraction. Dark matter is estimated to make up about 85 percent of the matter of the universe, though recent measurements of our own galaxy suggest that it may be as much as 90 percent of the universe. When an electron and a positron join to make a particle of dark matter, they give off 1.022 *MeV*, most commonly in the form of two photons of 0.511 *MeV* each. We see the signals of dark matter, as it is created from an electron, a positron, and two neutrinos, when we measure gamma rays coming from the dark matter halo surrounding our galaxy. This is the result of an event that we don't see, a collision of a powerful photon, or perhaps a relativistic electron, with a particle of dark matter, splitting it apart.

Dark matter is an equal mix of matter and antimatter. If this description of dark matter is correct, then the matter-antimatter imbalance isn't as bad as we thought. The spins of the electron and positron in dark matter may be aligned in the same direction. This is different than either paired electrons in atoms or paired electrons in Cooper pairs, where the spins are perpendicular. In SUiTe, the alignment of spins in the same direction can only be done in rigid particles, and requires the two electrons that are on separate ribs in a particle of dark matter to be very close to each other. This arrangement would help to provide internal stability for the structure of dark matter. If the spins of dark matter particles are arranged in this way, they will be spin one-half particles like electrons despite their construction from an even number of spin one-half particles.

A universe must have space, matter, mass, energy, and time. SUiTe defines space and matter. Mass is the interaction of space with matter and can be described as either rest mass or inertial mass. Rest mass does not change, but inertial mass changes with changes to the density of space surrounding the particle of matter. Energy is due to forces acting on matter. Energy, like inertial mass, can change with changes to the density of space. Time is more difficult to explain.

The direction of time in entropy is due to relative probabilities. Forward time is nothing more than a path that is more likely in terms of probability than the path that we would regard as being backward in time. The velocity of an object is encoded in the tilt of its forcelines, but we are unable to see or measure this tilt. In order to know the velocity of an object, we will need to have time. Our measurement of time, in this case, must also be the result of the movement of an object, a repetitive movement where each repetition is a fraction of time.

For the multiverse to exist, the fundamental particles must have existed. The creation of space by the fundamental particles coincides with the beginning of time. This is not the form of time that can be measured. That form of time requires matter and repetitive motion. The initial time has a beginning and an end. It is either on or off. The multiverse either exists or it doesn't.

Index

www.ingramcontent.com/pod-product-compliance
Lightning Source LLC
LaVergne TN
LVHW052302200926
848481LV00002B/437